Lecture Notes in Mathematics

A collection of informal reports and seminars
Edited by A. Dold, Heidelberg and B. Eckmann, Zürich

329

Walter Trebels
Lehrstuhl A für Mathematik der Technischen Hochschule Aachen,
Aachen/BRD

Multipliers for (C, α)-Bounded Fourier Expansions in Banach Spaces and Approximation Theory

Springer-Verlag
Berlin · Heidelberg · New York 1973

AMS Subject Classifications (1970): Primary: 40J05, 41A25
Secondary: 26A33, 40D15, 40G05, 42A56

ISBN 3-540-06357-9 Springer-Verlag Berlin · Heidelberg · New York
ISBN 0-387-06357-9 Springer-Verlag New York · Heidelberg · Berlin

Offsetdruck: Julius Beltz, Hemsbach/Bergstr.

PREFACE

In recent years some of the fundamental problems of approximation theory have turned out to be the verification of Jackson-, Bernstein-, and Zamansky-type inequalities for particular approximation processes, a study of the comparison of two different processes with respect to their rate of convergence, as well as the associated problems of non-optimal and optimal (or saturated) approximation for given processes.

These problems are here examined in the frame of abstract Fourier series in Banach spaces with respect to a total, fundamental sequence of mutually orthogonal projections $\{P_k\}$, the approximation processes being of multiplier type - i.e. (in some sense) summation methods of the abstract series.

In view of the multiplier structure it turns out that problems such as those mentioned above may be transferred to corresponding ones upon the coefficients (associated to the approximation processes in question) in the form of uniform multiplier conditions. In order to check such conditions multiplier criteria are required. To develop such, by the applications in mind it is most convenient to assume the uniform boundedness of the Cesàro means of order α (for some $\alpha \geq 0$) of the abstract expansion. Fractional α are admitted since the applications are to cover, for example, the Riesz means of classical trigonometric series expansions. Now the mere hypothesis that the (C,α)- means are bounded suffices to develop a most useful multiplier theory for the Banach space X with respect to the projections $\{P_k\}$.

In particular, it turns out that each scalar sequence $\eta = \{\eta_k\}_{k\in \mathbb{P}}$ ($\mathbb{P}$ the set of non-negative integers) belonging to the set $bv_{\alpha+1}$ is a multiplier sequence; here the $bv_{\alpha+1}$-norm of η is some sum of (fractional) differences of η. Since, in general, it is difficult to check whether $\eta \in bv_{\alpha+1}$ or not, η on $\mathbb{P}$ is suitably extended to a function $e(x)$ defined on $\mathbb{R}_1^+$ belonging to a suitable set $BV_{\alpha+1}$ (BV_1 is the set of functions of bounded variation). Then one has the fundamental inclusions

$$(*) \qquad BV_{\alpha+1}|_{\mathbb{P}} \subset bv_{\alpha+1} \subset M.$$

The methods employed involve the theories of fractional differences and differentiation, many results of which the author first had to polish up. In particular, results of classical summability theory such as those of A.F. Andersen 1928 (in connection with the set $bv_{\alpha+1}$) and of G.H. Hardy 1916 (Second theorem of consistency) were suitably modified. With respect to the theory of fractional differentiation, results of H. Weyl 1917 (Lipschitz conditions and the existence of fractional derivatives), J.J. Gergen 1937 (connection between sums over fractional differences and integrals over fractional derivatives) as well as J. Cossar 1941 (definition of a fractional derivative) could be used or carried over.

The multiplier theory is so sharp that relation (*) is necessary and sufficient for the Cesàro means of the same order to be uniformly bounded. On the other hand, this multiplier theory is quite useful (also in the strictly fractional case) as is shown by several examples such as the means of Abel-Cartwright, Riesz, Picard, Cesàro, de La Vallée-Poussin. For these approximation processes the series of problems mentioned above is solved in Banach spaces, provided only that the expansions in X with respect to $\{P_k\}$ have uniformly bounded Cesàro means of some order.

Finally, particular choices of the Banach space and its sequence of projections yield new and deep applications to one - and multiple trigonometric series, to Laguerre and Hermite series, as well as to expansions into Jacobi polynomials or spherical harmonics. For all these expansions (and those which are (C,α)-bounded) this elegant and lucid approach gives a multiplier theory together with a large number of new approximation theoretical results and reveals the outstanding role of Cesàro (or Riesz) summability.

Summarizing, let me emphasize that one of the major contributions of Trebels' work is the fact that he seems to have been the first to apply the theory of divergent series (in its vector-valued form) - a theory consisting of "hard analysis" that took about three centuries to develop - to the more modern approximation theory. Furthermore, he clearly pointed out an intimate connection (still to be established in its full generality) between summability theory of abstract orthogonal expansions and multiplier theory. Sharper results in summability theory should lead to sharper results in multiplier theory (and conversely), and these in turn will lead to a sharper approximation theory - a most promising and interesting research problem. However, concerning (C,α)-bounded expansions, this problem has been solved and is to be found in this contribution. It also delivers in a unified presentation very practical multiplier theories for Laguerre, Hermite and Jacobi series, which seem to be new. The present article, which has been written in form of a monograph, should receive a wide audience.

Aachen, January 1973

P.L. Butzer

CONTENTS

1. INTRODUCTION

1.1 General background

The origin of the present investigation is to be seen in two papers by Butzer - Nessel - Trebels [33; I,II] concerned with two problems of Favard [43, 44], namely on the comparison of summation processes of abstract Fourier series in a Banach space X (with respect to their rate of convergence) and on the saturation problem. Whilst treating these problems it turned out that there would be needed a suitable multiplier theory. In order to develop one, and in view of the applications in mind (to Fourier, ultraspherical, Laguerre series, and so on), it was convenient to assume the Cesàro means of order j ($j \in \mathbf{P}$, the set of non-negative integers) of the abstract Fourier expansion to be uniformly bounded. The sufficient multiplier criterium resulting from this is the classical one for numerical series summable (C,j) (cf.[61; p.128]).

The parallelism existing between multiplier criteria for numerical (C,α)-bounded series ($\alpha \geq 0$) and for classical Fourier series has already been observed by Moore [75] and Goes [53, 54] in their investigations on sufficient multiplier criteria for one-dimensional, trigonometric series expansions on several particular function spaces with the aid of suitable sums of fractional differences; for further literature see [52].

Apart from the abstract framework, the following idea (cf.[33]) is a decisive advantage with respect to applications, namely to extend suitably the multiplier sequence $\eta = \{\eta_k\}$, defined on $\mathbf{P}$, to a function $e(x)$ defined on the positive half-axis ($x \geq 0$) and then estimate the sum (over the differences of η) by a suitable integral over derivatives of the function $e(x)$. (By the way, this procedure is already implicitly contained in [61; p.373]).

However, though in [33] α is restricted to $\mathbf{P}$ the example of the Riesz-means (see (3.35)) of classical trigonometric series expansions (for small exponents λ) calls in [33] for an extension of the theory to all $\alpha \geq 0$. This will be performed in Section 3, where an analog of Hardy's [60] "Second theorem of consistency" is added, which will considerably simplify the computations in the applications. Let us remark that the resulting integral criterium, i.e., the extension of the above mentioned criterium (cf.[61; p.373]) to all $\alpha \geq 0$ is almost identical to criteria for numerical series by Borwein [20], Maddox [71], Russel [86] and others, though their context as well as their methods seem to be quite different.

Our approach is based immediately upon the hypothesis that the (C,α)-means of the abstract expansion is uniformly bounded. We essentially use Gergen's [50] elegant proof on the equivalence of Cesàro and Riesz summability of numerical series which can be carried over to our situation without difficulties. (The corresponding proof of Ingham [65] will probably also work; for α an integer this is obvious by [61; p.113]).

Since many results on (C,α)-boundedness in norm of concrete orthogonal expansions have only been proved in the last ten years, there was no great demand for a unified approach to norm estimates in the past, and there seem to be only a few papers using the parallelism between the multiplier theory of classical summability theory and of expansions in Banach spaces (one of the first seems to have been Hille [63]). On the other hand, for pointwise results this correspondence has been extensively used in case of Dirichlet, Fourier, power series, and so on (see e.g. the books [39; Ch.3,4], [61], [108; p.154], and only recently [97; p.285]).

A systematic exploitation of this parallelism will give a number of analogous results for various Banach-valued expansions, and not only (C,α)-bounded ones, as well as results for operators from X into another space (for one-dimensional trigonometric series see e.g. [53, 54]).

But here we restrict ourselves to multiplier theory for (C,α)-bounded, Banach-valued expansions in X, and to its application to several (norm-) approximation problems.

Before sketching the latter ones let us mention the "transplantation" approach as given axiomatically for series expansions by Gilbert [51] which generalizes basic work of Askey-Wainger [8] and Askey [4]. This approach is based heavily upon Riesz' theorem that the partial sums (i.e. $(C,0)$-means) of the classical one-dimensional Fourier series converge in norm for $1 < p < \infty$. It allows one to associate the trigonometric series with an expansion in functions $\{u_k\}$ defined on $(0,\pi)$ (not necessarily orthogonal), which are "similar" to $\{\cos kx\}$ and {sin kx} (e.g. perturbed cosine and sine functions, Fourier-Bessel and Fourier-Dini functions, Jacobi polynomials, eigenfunctions of fairly general Sturm-Liouville problems and so on - see [51]). Then the Abel means of the latter expansion are uniformly bounded in L^p; using the Marcinkiewicz multiplier theorem for the trigonometric system in weighted L^p-spaces one can obtain its analog for the system $\{u_k\}$ (further restrictions on the domain of p seem to be involved, see [8], [4]).

The advantages of the "transplantation" approach are the following: i) one does not need any knowledge on (C,α)-boundedness of the expansion, ii) the system need not necessarily be orthogonal, and iii) the sharp known results for the trigonometric system may be applied.

The advantages of our approach via the (C,α)-means are to be seen in: i) it is not restricted to functions similar to {cos kx}, {sin kx}, ii) multiplier criteria may be derived in an elementary and direct manner on all L^p-spaces from the (C,α)-boundedness hypothesis.

Now, for the sake of completeness, let us briefly sketch the approximation problems for which we wish to apply the multiplier theory mentioned above.

1.2 Approximation theory in Banach spaces

Generalizing classical approximation theorems of Jackson, Bernstein, Zamansky, etc., Butzer-Scherer [37] have shown that in any Banach space X there holds a general approximation theorem for linear approximation processes (with respect to their rate of convergence) provided only that Jackson- and Bernstein-type inequalities are satisfied for these processes.

To give a few details concerning this theory let X be a Banach space with norm $\|\cdot\|$, and let $[X]$ be the set of all bounded linear operators from X into X. Consider a family $\mathcal{T} = \{T(\varepsilon);\ \varepsilon \in [0,1]\}$ of strongly measurable operators in $[X]$ satisfying

$$(1.1)\qquad T(0) = I,\quad T(\varepsilon_1)T(\varepsilon_2) = T(\varepsilon_2)T(\varepsilon_1) \qquad (\varepsilon_1,\varepsilon_2 \in [0,1]),$$

$$(1.2)\quad \|T(\varepsilon)f\| \leq M_{\mathcal{T}}\|f\|, \quad \lim_{\varepsilon\to 0+}\|T(\varepsilon)f-f\| = 0 \qquad (f \in X);$$

further introduce a Banach subspace Y of X with semi-norm $|\cdot|_Y$ and norm $\|\cdot\|_Y = |\cdot|_Y + \|\cdot\|$ so that Y is continuously embedded in X, in notation $Y \subset X$. Then

Definition 1.1. *Let $Y \subset X$ and $y(\varepsilon)$ be a monotonely increasing function such that $0 < y(\varepsilon) \leq y(1) = 1$ and (m_y being a constant)*

$$(1.3)\qquad y(\varepsilon) \leq m_y\, y(\varepsilon/2) \qquad (\varepsilon \in (0,1]).$$

a) $\mathcal{T}$ is said to satisfy a Jackson-type inequality of order $y(\varepsilon)$ on X with respect to Y provided

$$\|T(\varepsilon)f-f\| \leq D_Y\, y(\varepsilon)|f|_Y \qquad (f \in Y)$$

for some constant D_Y (independent of f and ε).

b) $\mathcal{T}$ is said to satisfy a Bernstein-type inequality of order $y(\varepsilon)$ on X with respect to Y provided $T(\varepsilon)$ is strongly measurable on Y (in particular $T(\varepsilon)(X) \subset Y$) and

$$|T(\varepsilon)f|_Y \leq D_Y^*[y(\varepsilon)]^{-1}\|f\| \qquad (f \in X)$$

for some constant D_Y^*.

Now we may formulate a quite special case of the general approximation theorem of Butzer-Scherer [37; Cor.2] mentioned above:

Theorem 1.2. *Let $Y \subset Z \subset X$ and $\mathcal{T}$ be as above satisfying Jackson- as well as Bernstein-type inequalities of orders $y(\varepsilon)$ and $z(\varepsilon)$ on X with respect to Y and Z, respectively. Let $y(\varepsilon)$ and $z(\varepsilon)$ satisfy*

$$(1.4)\qquad \begin{aligned} &\int_0^\varepsilon y(u)[z(u)]^{-1}\, u^{-1}du = O\big(y(\varepsilon)[z(\varepsilon)]^{-1}\big) \\ &\int_\varepsilon^1 [y(u)]^{-1}\, z(u)\, u^{-1}du = O\big([y(\varepsilon)]^{-1}\, z(\varepsilon)\big). \end{aligned}$$

\- *Let $\Omega(\varepsilon)$ be a positive, nondecreasing function satisfying*

$$(1.5)\quad \begin{aligned} &\int_0^{\varepsilon} [z(u)]^{-1}\,\Omega(u)u^{-1}du = O\big([z(\varepsilon)]^{-1}\Omega(\varepsilon)\big),\\ &\int_{\varepsilon}^{1} [y(u)]^{-1}\,\Omega(u)u^{-1}du = O\big([y(\varepsilon)]^{-1}\Omega(\varepsilon)\big). \end{aligned}$$

Then the following assertions are equivalent for $\varepsilon\to 0+$:

(a) $\quad \|T(\varepsilon)f-f\| = O(\Omega(\varepsilon))$,

(b) $\quad |T(\varepsilon)f|_Y = O\big([y(\varepsilon)]^{-1}\Omega(\varepsilon)\big)$,

(c) $\quad f\in Z,\ |T(\varepsilon)f-f|_Z = O\big([z(\varepsilon)]^{-1}\Omega(\varepsilon)\big)$,

(d) $\quad K(y(\varepsilon),f;\ X,Y) \equiv \inf_{g\in Y}\big(\|f-g\| + y(\varepsilon)|f|_Y\big) = O(\Omega(\varepsilon))$.

If in addition Ω satisfies the further conditions

$$(1.6)\quad \begin{aligned} &\int_{\varepsilon}^{1} z(u)[\Omega(u)]^{-1}u^{-1}du = O\big(z(\varepsilon)[\Omega(\varepsilon)]^{-1}\big),\\ &\int_0^{\varepsilon} y(u)[\Omega(u)]^{-1}u^{-1}du = O\big(y(\varepsilon)[\Omega(\varepsilon)]^{-1}\big), \end{aligned}$$

then the following assertions are equivalent for $1\leqslant q<\infty$:

(a)' $\quad \int_0^1 \{[\Omega(u)]^{-1}\|T(u)f-f\|\}^q\,u^{-1}du < \infty$,

(b)' $\quad \int_0^1 \{[\Omega(u)]^{-1}\,y(u)|T(u)f|_Y\}^q\,u^{-1}du < \infty$,

(c)' $\quad f\in Z,\ \int_0^1 \{[\Omega(u)]^{-1}\,z(u)|T(u)f-f|_Z\}^q\,u^{-1}du < \infty$,

(d)' $\quad \int_0^1 \{[\Omega(u)]^{-1}\,K(y(u),\ f;X,Y)\}^q\,u^{-1}du < \infty$.

Let us remark that for the particular choice $\Omega(\varepsilon) = \varepsilon^{\beta}$, $y(\varepsilon) = \varepsilon^{m}$, and $z(\varepsilon) = \varepsilon^{k}$, the assumptions read $0 \leqslant k < \beta < m$, and that in concrete spaces the K-functional (introduced by Peetre, cf.(d)) corresponds to a modulus of continuity (cf. [26; Ch.4]). For more general results we refer to [37], for applications to [35,36,37]. Let us only mention that there is also a discrete version (instead of $\{T(\varepsilon);\ 0 \leq \varepsilon \leq 1\}$ one has a countable family $\{S(n)\} \subset [X]$ satisfying (1.1) and (1.2)), in which case the measurability conditions simply reduce to $S(n)f \in Y$ and $S(n)f \in Z$, respectively. The choice of $0 \leq \varepsilon \leq 1$ is only technical and may be replaced by $\varepsilon^{-1} = \rho$, $\rho \geq 1$. Clearly one may take $\rho > 0$, and since this notation coincides with the standard one, we will use it henceforward.

For an application of Theorem 1.2 one essentially has to check whether convenient Jackson- and Bernstein-type inequalities are satisfied. This will be carried out in Sections 2, 4, 5 in connection with summation processes of Fourier expansions in Banach spaces (without reformulating Theorem 1.2 in concrete examples).

The saturation problem, mentioned at the beginning, may be interpreted as the problem of an optimal Jackson-type inequality. It was first introduced by Favard for summation methods of trigonometric series in a lecture in 1947 (cf.[43]) and may be formulated as follows (see e.g. [31; p.434]).

Definition 1.3. *The strong approximation process* $\mathcal{T} = \{T(\rho);\ \rho > 0\}$ *(cf. (1.2)) is said to possess the saturation property if there exists a positive function* $\Theta(\rho)$, $\rho > 0$, *tending monotonely to infinity as* $\rho \to \infty$ *such that every* $f \in X$ *for which*

$$\|\Theta(\rho)[T(\rho)f-f]\| = o(1) \qquad (\rho \to \infty)$$

is an invariant element of $\mathcal{T}$, *i.e.* $T(\rho)f = f$ *for all* $\rho > 0$, *and if the set*

$$F[X;\mathcal{T}] = \{f \in X;\ \|\Theta(\rho)[T(\rho)f-f]\| = O(1),\ \rho \to \infty\}$$

contains at least one noninvariant element. In this event, the approximation process $\mathcal{T}$ *is said to have optimal approximation order* $[\Theta(\rho)]^{-1}$ *or to be saturated in* X *with order* $[\Theta(\rho)]^{-1}$, *and* $F[X;\mathcal{T}]$ *is called its Favard or saturation class.*

Today there exists a vast literature concerned with saturation for various types of approximation processes. To mention general approaches in regard to solution, there exists an integral transform method in diverse Lebesgue spaces as well as the semi-group method on arbitrary Banach spaces in its extended form (for detailed bibliographical comments one may consult the books of Berens [15], Butzer-Berens [26], and Butzer-Nessel [31]).

The implication (a) $\Rightarrow$ (b) in Theorem 1.2 is called a Zamansky-type inequality which, however, suffers under the restrictions (1.5) and (1.6) upon Ω. In case $[y(\varepsilon)]^{-1} = \Theta(\rho)$ is the saturation order and the relative completion of Y is $F[X;\mathcal{T}]$ (see Def. 2.5), this inequality may be established without the above restrictions via the direct estimate

$$(1.7) \qquad |T(\rho)f|_Y \leq D\,\Theta(\rho)\|T(\rho)f-f\| \qquad (f \in X).$$

These matters, as well as extensions, will be treated in Sections 2, 4, 5 for approximation processes given via Fourier expansions in Banach spaces.

Let us finally introduce the comparison problem for two summation methods, mentioned at the beginning and posed by Favard [44].

Definition 1.4. *Let $\mathcal{S}$ and $\mathcal{T}$ be two approximation processes on X satisfying (1.2). $\mathcal{T}$ is said to be better than $\mathcal{S}$ (with respect to its rate of convergence), if there exists a constant $D > 0$ such that*

$$(1.8) \qquad \|T(\rho)f-f\| \leq D\|S(\rho)f-f\| \qquad (f \in X;\ \rho > 0).$$

If $\mathcal{T}$ is better than $\mathcal{S}$ and the latter in turn better than $\mathcal{T}$, then the processes are said to be equivalent, in notation

$$\|T(\rho)f-f\| \approx \|S(\rho)f-f\| \qquad (f \in X).$$

First contributions to this problem have been made by Shapiro [89], Boman-Shapiro [19], and Butzer-Nessel-Trebels [32, 33;I] (compare the comments in [31; p.507], [32]). Whereas in [89], [19] (cf. Löfström [70] for precursory material) the concrete case of approximation processes representable as Fourier convolution integrals of Fejér's type is considered for Euclidean n-space (or n-dimensional torus), in [32, 33;I] the problem is discussed in the setting of abstract Hilbert spaces and of expansions in Banach spaces, respectively.

We dispense with a survey of the following sections and refer to the short summary preceding each section.

Acknowledgements

These investigations were carried out under a "DFG-Habilitanden-stipendium" and the author is much obliged to the DFG for its generous support. He is very much indebted to Professor P.L. Butzer for his continuous encouragement and promotion. It was he who suggested the problem which was developped in cooperation with him and Professor R.J. Nessel; the two papers [33; I,II] reveal their decisive contributions; for this and their critical reading he is profoundly grateful. He also acknowledges with pleasure the numerous helps and hints given in various discussions by his colleagues, in particular by Doz. Dr. E. Görlich and Drs. H. Johnen, K. Scherer, E.L. Stark and U. Westphal

2. GENERAL THEORY

In this section the concept of abstract expansions in Banach spaces X is introduced with the aid of a countable family of projections $\{P_k\}$, and multipliers with respect to a fixed pair $X, \{P_k\}$ are studied. Restricting the approximation processes to those of multiplier type (i.e. some kind of "convolution" structure is assumed), sufficient conditions for Jackson-, Bernstein-, Zamansky-type inequalities as well as for saturation and comparison theorems are formulated in terms of multipliers. The framework as well as the comparison theorem is taken over from [33;I]; for the saturation theorem see [33;II], for the Bernstein-type inequality Görlich-Nessel-Trebels [59].

2.1 Notations and further definitions

As in Sec. 1.2, let X be an arbitrary (real or complex) Banach space with norm $\|\cdot\|$ and elements f, g,...; let [X] be the Banach algebra of all bounded linear operators on X into itself. Further, denote by $\mathbb{R}$ and $\mathbb{C}$ the set of all real and complex numbers, respectively, and let $\mathbb{Z}$, $\mathbb{P}$, $\mathbb{N}$ be the sets of all, of all non-negative, of all positive integers, respectively. By $[\alpha]$ denote the largest integer less than or equal to $\alpha \in \mathbb{R}$. Let us decompose the Banach space X by a sequence of projections $\{P_k\}_{k\in\mathbb{P}} \subset [X]$ satisfying the following properties

i) the projections P_k are mutually orthogonal, i.e., for all $j, k \in \mathbb{P}$ there holds $P_jP_k = \delta_{j,k}P_k$, $\delta_{j,k}$ being Kronecker's symbol;

ii) the sequence $\{P_k\}$ is total, i.e., $P_kf = 0$ for all $k \in \mathbb{P}$ implies $f = 0$;

iii) the sequence $\{P_k\}$ is fundamental[1], i.e., the linear span of the ranges $P_k(X)$, $k \in \mathbb{P}$, is dense in X: $\overline{\bigcup_{k\in\mathbb{P}} P_k(X)} = X$.

---

1) On account of the Banach-Steinhaus theorem this property is a necessary one for a uniformly bounded family of operators $\{T(\rho)\}$ to imply convergence on all X provided $T(\rho)$ converges on each $P_k(X)$, $k \in \mathbb{P}$. However, it is irrelevant for the multiplier criteria in Sec.3.

Then with each $f \in X$ one may associate its (formal) Fourier series expansion

$$f \sim \sum_{k=0}^{\infty} P_k f \qquad (f \in X). \tag{2.1}$$

With s the set of all sequences $\eta = \{\eta_k\}_{k \in \mathbb{P}}$ of scalars, $\eta \in s$ is called a multiplier for X (corresponding to $\{P_k\}$), if for each $f \in X$ there exists an element $f^\eta \in X$ such that $\eta_k P_k f = P_k f^\eta$ for all $k \in \mathbb{P}$, thus

$$f^\eta \sim \sum_{k=0}^{\infty} \eta_k P_k f \, . \tag{2.2}$$

Note that f^η is uniquely determined by f since $\{P_k\}$ is total. The set of all multipliers is denoted by $M = M(X;\{P_k\})$. With the natural vector operations, coordinatewise multiplication and norm

$$\|\eta\|_M = \sup\{\|f^\eta\| ; \; f \in X, \; \|f\| \leq 1\}, \tag{2.3}$$

M is a commutative Banach algebra containing the identity $\{1\} \in s$.

An operator T from X into itself is called a multiplier operator if there exists a sequence $\tau \in s$ such that $P_k T f = \tau_k P_k f$ for all $f \in X$, $k \in \mathbb{P}$, i.e., one has the formal expansion

$$Tf \sim \sum_{k=0}^{\infty} \tau_k P_k f \qquad (f \in X). \tag{2.4}$$

Obviously, $T \in [X]$. Thus, by definition, with each multiplier operator $T \in [X]_M$ (the set of all multiplier operators on X) there is associated a multiplier sequence $\tau \in M$ and vice versa, and since $\|T\|_{[X]} = \|\tau\|_M$ by definition (cf. (2.3)), M can be identified with $[X]_M$. In the future we always assume $\Gamma, \mathcal{T} \subset [X]_M$.

<u>Remark.</u> The expansion (2.1) represents a slight generalization of the concept of Fourier series in a Banach space X associated with a fundamental, total, biorthogonal system $\{f_k, f_k^*\}$. Here $\{f_k, f_k^*\}$ consists of two sequences $\{f_k\} \subset X, \{f_k^*\} \subset X^*$ such that i) $f_j^*(f_k) = \delta_{j,k}$ for all $j,k \in \mathbb{P}$ (orthogonal), ii) $f_k^*(f) = 0$ for all $k \in \mathbb{P}$ implies $f = 0$ (total), and iii) the linear span of $\{f_k\}$ is dense in X (fundamental).

Then (2.1) and (2.4) read

$$f \sim \sum_{k=0}^{\infty} f_k^*(f)f_k, \quad Tf \sim \sum_{k=0}^{\infty} \tau_k f_k^*(f)f_k ,$$

respectively.

For these definitions and results compare Marti [72;p.86 ff], Milman [73] , Singer [91;pp.1-49] , etc.

In this framework, the general approximation Theorem 1.2 of Butzer-Scherer [37] suggests that one determines subspaces of X via some sequences of s which do not necessarily belong to M. For arbitrary $\psi \in s$ we define

$$(2.5) \qquad X^\psi = \{f \in X; \exists\, f^\psi \in X \text{ with } \psi_k P_k f = P_k f^\psi \text{ for all } k \in \mathbb{P}\} .$$

Obviously, if B^ψ is the operator with domain X^ψ and range in X defined by $B^\psi f = f^\psi$, $f \in X^\psi$, then B^ψ is a closed linear operator for each $\psi \in s$. Since $P_k(X)$ is contained in X^ψ for each $k \in \mathbb{P}$, B^ψ is densely defined. Further, defining a semi-norm on X^ψ via $|f|_\psi = \|B^\psi f\|$, X^ψ is a Banach space with respect to the norm $\|f\| + |f|_\psi$, and $X^\psi \subset X$.

2.2 Jackson- and Bernstein-type inequalities

The first general result, in fact just a reformulation in the present setting, reads (cf. [33, 59])

Theorem 2.1. *Let $\mathcal{T} \subset [X]_M$ be a strong approximation process with associated multiplier family $\{\tau(\rho)\}_{\rho>0}$.*

a) If there exist a non-negative, monotonely increasing function $\chi(\rho)$ with $\lim_{\rho\to\infty}\chi(\rho) = \infty$, $\psi \in s$, and a uniformly bounded multiplier family $\{\eta(\rho)\} \subset M$ with

$$(2.6) \qquad \chi(\rho)\{\tau_k(\rho) - 1\} = \psi_k \eta_k(\rho),$$

then one has the Jackson-type inequality

(2.7) $$\chi(\rho)\|T(\rho)f-f\| \leq \sup_{\rho>0} \|\eta(\rho)\|_M |f|_\psi \qquad (f \in X^\psi).$$

b) If (2.6) is replaced in a) by

(2.8) $$\psi_k \tau_k(\rho) = \chi(\rho)\eta_k(\rho) ,$$

then there holds the Bernstein-type inequality

(2.9) $$|T(\rho)f|_\psi \leq \chi(\rho) \sup_{\rho>0} \|\eta(\rho)\|_M \|f\| \qquad (f \in X).$$

The proofs of a) and b) immediately follow by the hypotheses. For example, since $\{P_k\}$ is total, (2.6) is equivalent to

$$\chi(\rho)\{T(\rho)f-f\} = E(\rho)(B^\psi f),$$

where $E(\rho)$ is the (uniformly bounded) operator associated to the (uniformly bounded) multiplier $\eta(\rho)$; hence (2.7) holds.

This theorem induces one to expect that the verification of multiplier conditions (such as (2.6) and (2.8)) will present the actual problem, and Section 3 is therefore devoted to establishing convenient criteria concerning multipliers.

2.3 A saturation theorem

Let $\mathsf{K} = \{k \in \mathsf{P} \; ; \; \tau_k(\rho) = 1$ for all $\rho > 0\}$ and assume $\mathsf{K} \neq \mathsf{P}$. Then the following condition upon $\mathcal{T}$ ensures the saturation property.

Definition 2.2. *The approximation process $\mathcal{T} \subset [X]_M$ satisfies condition (F), if there exist $\varphi \in s$ with $\varphi_k \neq 0$ for $k \notin \mathsf{K}$ and a non-negative, monotonely increasing function $\theta(\rho)$ with $\lim_{\rho\to\infty} \theta(\rho) = \infty$ such that*

(2.10) $$\lim_{\rho\to\infty} \theta(\rho)\{\tau_k(\rho) - 1\} = \varphi_k \qquad (k \in \mathsf{P})$$

Condition[2] (F) is a standard one in the study of saturation for summation processes of trigonometric series (cf. [31;p.435]). In fact, it was already introduced by Favard [43] in connection with fundamental, total biorthogonal systems (cf. Remark in Sec. 2.1) in arbitrary Banach spaces. As a consequence, the following result is substantially contained in [43] (the formulation is taken over precisely from [33;II]).

Lemma 2.3. *Let $f \in X$ and $\mathcal{T}$ satisfy condition (F).*

a) If there exists $g \in X$ such that

$$\lim_{\rho\to\infty} \|\Theta(\rho)\{T(\rho)f-f\} - g\| = 0 ,$$

the Fourier expansion of g is given by $g \sim \sum_{k=0}^{\infty} \varphi_k P_k f$.

b) $\Theta(\rho)\|T(\rho)f-f\| = o(1)$ implies $f \in \bigcup_{k\in K} P_k(X)$, and $T(\rho)f = f$ for all $\rho > 0$, thus f is an invariant element.

c) There exists some noninvariant $h \in X$ with $\Theta(\rho)\|T(\rho)h-h\| = O(1)$.

Proof. a) Since $P_k \in [X]$ and

$$P_k(\Theta(\rho)\{T(\rho)f-f\}) = \Theta(\rho)\{\tau_k(\rho)-1\} P_k f$$

one has for each $k \in \mathbb{P}$

$$\begin{aligned}\|\varphi_k P_k f - P_k g\| &= \lim_{\rho\to\infty} \|\Theta(\rho)\{\tau_k(\rho) - 1\}P_k f - P_k g\| \\ &\leq \lim_{\rho\to\infty} \|P_k\|_{[X]} \|\Theta(\rho)\{T(\rho)f-f\} - g\| = 0 ,\end{aligned}$$

which proves the assertion.

b) With $g = 0$ part a) gives $\varphi_k P_k f = 0$ for all $k \in \mathbb{P}$. In case $k \notin K$ it follows that $P_k f = 0$, whereas for $k \in K$ the normalization $\tau_k(\rho) = 1$

2) Note that in (2.6) - (2.9) the choice of ψ and χ is variable, whereas φ and Θ in (2.10) and (2.11) are determined by the process.

for all $\rho > 0$ gives $P_k T(\rho)f = P_k f$. Thus $P_k T(\rho)f = P_k f$ for all $k \in \mathsf{P}$, and since $\{P_k\}$ is total, the assertion follows.

c) Since for any $h \in P_k(X)$

$$\|T(\rho)h-h\| = |\tau_k(\rho) - 1|\,\|h\|,$$

$h \neq 0$ is noninvariant if $k \notin \mathsf{K}$, and the assertion follows by condition (F).

Definition 2.4. *The approximation process $\mathcal{T} \subset [X]_M$ is said to satisfy condition (F^*), if (F) holds and there exists a uniformly bounded multiplier family $\{\eta(\rho)\} \subset M$ such that*

$$\Theta(\rho)\{\tau_k(\rho) - 1\} = \eta_k(\rho)\psi_k \qquad (k \in \mathsf{P}, \rho > 0). \tag{2.11}$$

Condition (F^*) is also standard in saturation theory (cf. [31;Sec. 12.6] for detailed comments). Certainly, (F^*) (in connection with (F)) implies $\lim_{\rho\to\infty} \eta_k(\rho) = 1$ (it is assumed on K) so that by the theorem of Banach-Steinhaus the family $\{E(\rho)\}$ of operators corresponding to $\{\eta(\rho)\}$ forms a strong approximation process ($\{P_k\}$ is fundamental) satisfying $E(\rho)(X) \subset X^\psi$ for all $\rho > 0$. Relation (2.11) immediately implies

$$\|\Theta(\rho)\{T(\rho)f-f\}\| = \|B^\psi E(\rho)f\| \qquad (\rho > 0;\ f \in X) \tag{2.12}$$

and we have to discuss conditions upon f such that these expressions are uniformly bounded in ρ.

In this context, the idea of relative completion turns out to be fundamental (cf. Berens [15;p.14,p.28], [31;Sec.10.4]).

Definition 2.5. *Let $Y \subset X$ with semi-norm $|\cdot|_Y$. The completion of Y relative to X, denoted by $Y^{\sim X}$, is the set of those elements $f \in X$ for which there exists a sequence $\{f_n\} \subset Y$ and a constant $D > 0$ such that $|f_n|_Y \leq D$ for all n together with $\lim_{n\to\infty}\|f_n - f\| = 0$. With any $f \in Y^{\sim X}$ one may associate the semi-norm*

$$|f|_{Y^{\sim X}} = \inf\{\sup|f_n|_Y;\ \{f_n\} \subset Y,\ \lim_{n\to\infty}\|f_n - f\| = 0\}.$$

Note that $Y^{\sim X} = Y$ provided Y is reflexive. Using the notation $|f|_{(X^{\varphi})^{\sim X}} = |f|_{\varphi^{\sim}}$ one obtains

<u>*Theorem 2.6.*</u> *a) The following semi-norms are equivalent*[3] *on* $(X^{\varphi})^{\sim X}$:

i) $|f|_{\varphi^{\sim}}$, *ii)* $\sup_{\rho>0} \| B^{\varphi} S(\rho) f \|$,

where $\mathcal{S} = \{S(\rho); \rho > 0\} \subset [X]_M$ *is a further approximation process with* $S(\rho)(X) \subset X^{\varphi}$.

b) If $\mathcal{T}$ *satisfies condition* (F^*), *then the Favard class of* $\mathcal{T}$ *is* $(X^{\varphi})^{\sim X}$ *and*

iii) $\sup_{\rho>0} \| \Theta(\rho)\{T(\rho)f - f\} \|$

is a further equivalent semi-norm.

Proof. a) First assume that $|f|_{\varphi^{\sim}} < \infty$. Then, by definition, there exists a sequence $\{f_n\} \subset X^{\varphi}$ such that $|f_n|_{\varphi} \leq D$ uniformly for all n and $\lim_{n\to\infty} \| f_n - f \| = 0$. Since $B^{\varphi}S(\rho) \in [X]_M$ by the closed graph theorem, and since B^{φ}, $S(\rho)$ commute for each $\rho > 0$, one has

$$\| B^{\varphi} S(\rho) f \| = \lim_{n\to\infty} \| B^{\varphi} S(\rho) f_n \| = \lim_{n\to\infty} \| S(\rho) B^{\varphi} f_n \|$$

$$\leq \sup_{\rho>0} \| S(\rho) \|_{[X]} \sup_n |f_n|_{\varphi} .$$

However, the left-hand side is independent of the particular choice of the sequence $\{f_n\}$, whereas the right-hand side is independent of ρ. Therefore

$$\sup_{\rho>0} \| B^{\varphi} S(\rho) f \| \leq C |f|_{\varphi^{\sim}} ,$$

proving one direction of the assertion. The converse one is easily seen by examining the particular sequence $\{S(n)f\} \subset X^{\varphi}$.

---

3) Two semi-norms $|\cdot|_1$, $|\cdot|_2$ on Y are called equivalent: $|\cdot|_1 \sim |\cdot|_2$, if there exist constants $c_1, c_2 > 0$ such that $c_1|f|_1 \leq |f|_2 \leq c_2|f|_1$ for every $f \in Y$.

b) Since $\mathcal{E} = \{E(\rho);\rho > 0\}$ in condition (F^*) is an admissible choice for $\mathcal{T}$, the assertion immediately follows from (2.12).

Remark. Part a) is directly taken over from [33;II] . The theorem itself is a simple case of a saturation theorem of Berens [15;p.28] who, instead of (F^*), assumes the weaker Voronovskaja-type relation

$$\lim_{\rho\to\infty} \|\theta(\rho)\{T(\rho)f-f\} - Bf\| = 0 \tag{2.13}$$

with closed linear operator B, where $\mathcal{T}$, B are not necessarily of multiplier type. Under our hypotheses, (2.13) easily follows on account of (F^*) since $E(\rho)$ and B^φ commute on X^φ for each $\rho > 0$. Therefore, we do not formulate statements of type (2.13) explicitly. Let us mention that assertion $\|B^\varphi S(\rho)f\| = O(1)$ immediately meets standard representation theorems in case of the trigonometric system (cf. [31; p.233]). For characterizations of the present type in case of semigroups of operators one may consult [15;p.43] , [26;p.111] .

2.4 A Zamansky-type inequality; a comparison theorem

Zamansky-type inequalities (1.7) in terms of suitable multiplier conditions read as follows

Theorem 2.7. *Let $\mathcal{T} \subset [X]_M$ be a strong approximation process. If there exists a non-negative, monotonely increasing function $\chi(\rho)$ with $\lim_{\rho\to\infty} \chi(\rho) = \infty$, a sequence $\psi \in s$, and a uniformly bounded multiplier family $\{\eta(\rho)\}$ with*

$$\psi_k \tau_k(\rho) = \chi(\rho)\eta_k(\rho)(1 - \tau_k(\rho)) \qquad (k \in \mathbb{P};\rho > 0), \tag{2.14}$$

then one has the Zamansky-type inequality

$$\|B^\psi T(\rho)f\| \leq \chi(\rho)\Big(\sup_{\rho>0}\|\eta(\rho)\|_M\Big)\|T(\rho)f-f\| \qquad (f \in X). \tag{2.15}$$

Since the projections $\{P_k\}$ are total, the assertion is obvious.

Analogously one arrives at a comparison theorem (in the sense of Def. 1.4) for two summation methods $\mathcal{S}$ and $\mathcal{T}$ (see [33;I]).

<u>*Theorem 2.8.*</u> *Let $\mathcal{S}, \mathcal{T} \subset [X]_M$ be two approximation processes. Further, let there exist a uniformly bounded multiplier family $\{\eta(\rho)\}$ with*

$$\tau_k(\rho) - 1 = \eta_k(\rho)\{\zeta_k(\rho) - 1\} \qquad (k \in \mathbf{P}\ ; \rho > 0). \tag{2.16}$$

Then one has the following comparison result

$$\|T(\rho)f - f\| \le \left(\sup_{\rho>0}\|\eta(\rho)\|_M\right)\|S(\rho)f - f\| \qquad (f \in X). \tag{2.17}$$

The uniform multiplier conditions (2.6, 2.8, 2.11, 2.14, 2.16) are strong and intricate (clearly, the uniform boundedness of the multipliers may be relaxed to deliver weaker assertions); their verification in the applications is the actual problem. Therefore, the next section is devoted to establishing convenient criteria concerning (uniformly bounded) multipliers.

3. MULTIPLIER CRITERIA FOR (C,α)-BOUNDED EXPANSIONS

It seems hardly possible to develop a multiplier theory without imposing further hypotheses upon the projections $\{P_k\}$. A most convenient assumption is that the (C,α)-means of the expansion $f \sim \sum P_k f$ be uniformly bounded for some $\alpha \geq 0$. In accordance with classical summability theory it turns out that $\eta \in s$ is a multiplier on $(X,\{P_k\})$ if

$$(3.1) \qquad \sup_k |\eta_k| < \infty \quad \text{plus} \quad \sum_{k=0}^{\infty} A_k^{\alpha} |\Delta^{\alpha+1}\eta_k| < \infty \qquad (\alpha \geq 0)$$

(see (3.11) below). In Sec. 3.2 and 3.3 it will be shown that

$$(3.2) \qquad \sup_{x \geq 0} |e(x)| < \infty \quad \text{plus} \quad \int_0^{\infty} x^{\alpha} |de^{(\alpha)}(x)| < \infty \qquad (\alpha \geq 0)$$

is sufficient for (3.1) to hold, where $e(k) = \eta_k$ and $e^{(\alpha)}(x)$ is a suitable fractional derivative of order $\alpha \geq 0$ (see Cossar [41]). The analog of Hardy's [60] "Second Theorem of Consistency" to be established then yields, that (3.2) also holds for $e(\Phi(x))$ provided Φ satisfies several monotonicity and differentiability properties.

We separate the integral estimates in the integer case from those in the fractional case, since for integers these are simpler and more lucid than for fractionals. Nevertheless, let us emphasize that all estimates are based straight forwardly on the assumption that the (C,α)-means are uniformly bounded for some $\alpha \geq 0$ - a hypothesis always assumed in the following.

3.1 Classical multiplier criteria via differences

First let us introduce the (C,α)-means of f relative to $(X,\{P_k\})$ by

$$(3.3) \qquad (C,\alpha)_n f = \sum_{k=0}^{n} \{A_{n-k}^{\alpha} / A_n^{\alpha}\} P_k f \qquad (f \in X),$$

$$(3.4) \qquad A_n^{\alpha} = \binom{n+\alpha}{n} = \frac{\Gamma(n+\alpha+1)}{\Gamma(n+1)\Gamma(\alpha+1)} \qquad (\alpha \in \mathbb{R})$$

being the standard binominal coefficients. We observe for later use the formulae (cf. [2], [22], [110;I,p.77])

(3.5) $$\sum_{k=0}^{n} A_k^{\alpha} A_{n-k}^{\beta} = A_n^{\alpha+\beta+1} \qquad (\alpha,\beta \in \mathbb{R}, n \in \mathbb{P})$$

and

(3.6) $$A_n^{\alpha} = \frac{n^{\alpha}}{\Gamma(\alpha+1)} + O(n^{\alpha-1}) \qquad (-\alpha \notin \mathbb{N}).$$

As already announced, the (C,α)-boundedness of the expansion $\sum P_k f$ is presumed, i.e., for some $\alpha \geq 0$ there exists a constant $C_{\alpha} > 0$ such that for all $f \in X$

(3.7) $$\| (C,\alpha)_n f \| \leq C_{\alpha} \| f \| \qquad (n \in \mathbb{P}).$$

This hypothesis is always supposed in the following. In order to avoid long repetitions of the properties of $\{P_k\}$ in the subsequent results we introduce

Definition 3.0. The pair X, $\{P_k\}$ is said to satisfy the property (C^{α}) for some $\alpha \geq 0$ if X is a Banach space and $\{P_k\} \subset [X]$ is a total, fundamental system of mutually orthogonal projections satisfying (3.7) for some $\alpha \geq 0$.

If X, $\{P_k\}$ satisfies (C^{α}), then it follows by standard methods for all $\beta > \alpha$ that

(3.8) $$\| (C,\beta)_n f \| \leq C_{\alpha} \| f \| \qquad (n \in \mathbb{P}).$$

Another consequence is the following (cf. Theorem 2.6 b)

Theorem 3.1. Let X, $\{P_k\}$ satisfy condition (C^{α}) for some $\alpha \geq 0$. Suppose further that $\mathcal{T}$ satisfies condition (F) (see Def. 2.2) and that $\sup_{\rho>0} \Theta(\rho) \| T(\rho)f - f \|$ is finite. Then

$$\| \sum_{k=0}^{n} (A_{n-k}^{\alpha} / A_n^{\alpha}) \varphi_k P_k f \| \leq C_{\alpha} \sup_{\rho>0} \Theta(\rho) \| T(\rho)f - f \| .$$

Proof. By (F) one deduces that for each fixed $n \in \mathbb{P}$

$$\|\sum_{k=o}^{n} (A_{n-k}^{\alpha} / A_n^{\alpha})\varphi_k P_k f\|$$

$$= \lim_{\rho\to\infty} \| \sum_{k=o}^{n} (A_{n-k}^{\alpha} / A_n^{\alpha})\Theta(\rho)\{\tau_k(\rho)-1\}P_k f \|$$

$$\leq C_\alpha \sup_{\rho>o}\{\Theta(\rho)\| T(\rho)f-f\|\} .$$

Now let us derive a sufficient multiplier criterium on $(X;\{P_k\})$ (under the tacit hypothesis (3.7)). Analogous to the integer case in [33;I,II] consider for (fractional) $\alpha \geq 0$ the class

$$(3.9) \qquad bv_{\alpha+1} = \{\eta \in l^\infty;\| \eta\|_{bv_{\alpha+1}} = \sum_{k=o}^{\infty} A_k^{\alpha}|\Delta^{\alpha+1} \eta_k| + \lim_{k\to\infty} |\eta_k| < \infty\},$$

where l^∞ is the subset of s of all bounded sequences, the fractional difference operator Δ^β being defined according to Andersen [2] by

$$(3.10) \qquad \Delta^\beta \eta_k = \sum_{m=o}^{\infty} A_m^{-\beta-1} \eta_{k+m}$$

whenever the series on the right hand side converges in the ordinary sense; by (3.6) one clearly has absolute convergence provided $\beta \geq 0$ and $\eta \in l^\infty$. Obviously, this definition coincides with the usual one in the integer case; e.g. in case $\beta = 1$ one has $A_o^{-2} = 1$, $A_1^{-2} = -1$, and $A_m^{-2} = 0$, $m \geq 2$, on account of (3.4), and thus $\Delta \eta_k = \eta_k - \eta_{k+1}$. Turning towards the class $bv_{\alpha+1}$ we note that assumption $\lim_{k\to\infty} \eta_k = \eta_\infty$ is no real restriction, since the existence of the limit already follows from $\eta \in l^\infty$ and the boundedness of the sum in (3.9) (cf. [3]).

<u>*Lemma 3.2.*</u> *One has* $bv_{\alpha+1} \subset bv_{\gamma+1}$, $0 \leq \gamma < \alpha$. *For any* $\eta \in bv_{\alpha+1}$, $\alpha \geq 0$,

$$\eta_m - \eta_\infty = \sum_{k=o}^{\infty} A_k^{\alpha} \Delta^{\alpha+1}\eta_{k+m} \quad .$$

Proof. This lemma is essentially proved in Andersen [2] and the argument is repeated for the sake of completeness. First we observe that according to a formula of Andersen [2]

$$\Delta^\delta(\Delta^\alpha \eta_k) = \Delta^{\delta+\alpha}\eta_k \qquad (\delta \geq -1,\ \alpha > 0,\ \alpha+\delta \geq 0)$$

holds provided $\eta_k = o(1)$ for $k \to \infty$. Therefore consider the sequence $\{\eta_k - \eta_\infty\}_{k=o}^{\infty}$ and first choose $\alpha - 1 \leq \beta < \alpha$, $\delta = \alpha - \beta$. By (3.5) it is clear that $\sum_{k=0}^{\infty} A_k^{-\alpha+\delta-2} = 0$ and therefore $\Delta^{\alpha-\delta+1}\{\eta_k-\eta_\infty\} = \Delta^{\alpha-\delta+1}\eta_k$. Thus

$$\|\eta\|_{bv_{\beta+1}} - |\eta_\infty| = \sum_{k=0}^{\infty} A_k^{\alpha-\delta}|\Delta^{\alpha-\delta+1}\eta_k|$$

$$= \sum_{k=0}^{\infty} A_k^{\alpha-\delta}|\Delta^{-\delta}(\Delta^{\alpha+1}\eta_k)| \leq \sum_{k=0}^{\infty} A_k^{\alpha-\delta}\sum_{n=0}^{\infty} A_n^{\delta-1}|\Delta^{\alpha+1}\eta_{k+n}|$$

$$= \sum_{k=0}^{\infty} A_k^{\alpha}|\Delta^{\alpha+1}\eta_k| = \|\eta\|_{bv_{\alpha+1}} - |\eta_\infty| \quad .$$

Here it was used that the coefficient of $\Delta^{\alpha+1}\eta_m$ $(m \in P\ )$ in the double series is

$$\sum_{n=0}^{m} A_n^{\alpha-\delta}\, A_{m-n}^{\delta-1} = A_m^{\alpha}$$

by (3.5) and that the change of summation is justified by the absolute convergence (all terms are positive!). Analogously it follows that $\|\eta\|_{bv_\gamma} \leq \|\eta\|_{bv_{\alpha+1}}$ if $\max\{0,\alpha-2\} \leq \gamma < \alpha - 1$ so that the imbedding relation is established in case $\alpha - 2 \leq 0$. Otherwise a k-times iteration of this argument $(-1 \leq \alpha - 2 - k \leq 0)$ gives $bv_{\alpha+1} \subset bv_{\gamma+1}$.

Omitting the absolute value signs the above proof in particular gives

$$\sum_{k=0}^{\infty} A_k^{\beta}\, \Delta^{\beta+1}\{\eta_{k+m} - \eta_\infty\} = \sum_{k=0}^{\infty} A_k^{\alpha}\, \Delta^{\alpha+1}\eta_k \ ;$$

now $A_k^{\beta} = 1$, $k \in P$, in case $\beta = 0$ and therefore

$$\sum_{k=0}^{\infty} \Delta\{\eta_{k+m} - \eta_\infty\} = \eta_m - \eta_\infty \ ,$$

yielding the rest of the assertion.

With the aid of Lemma 3.2 we arrive at a first result concerning multipliers

Theorem 3.3. *Let X, $\{P_k\}$ satisfy condition (C^{α}) for some $\alpha \geq 0$. Then $bv_{\alpha+1}$ is continuously embedded in $M(X;\{P_k\})$, i.e.,*

$$\|\eta\|_M \leq C_\alpha \|\eta\|_{bv_{\alpha+1}} \qquad (\eta \in bv_{\alpha+1}) \ . \tag{3.11}$$

Proof (for $\alpha \in P$ cf. [33;I,II]). For each $f \in X$ and $\eta \in bv_{\alpha+1}$ set

$$f^{\eta} = \sum_{k=o}^{\infty} A_k^{\alpha}\ \Delta^{\alpha+1}\eta_k (C,\alpha)_k f + \eta_{\infty} f .$$

Then f^{η} exists in X since by (3.7) and (3.9)

$$\| f^{\eta}\| \leq C_{\alpha}\| f\| \sum_{k=o}^{\infty} A_k^{\alpha} |\Delta^{\alpha+1}\eta_k| + |\eta_{\infty}|\| f\| \leq C_{\alpha}\| \eta\|_{bv_{\alpha+1}}\| f\| .$$

Thus it remains to show that $f^{\eta} \sim \sum \eta_k P_k f$. Observing that

$$P_n(C,\alpha)_k f = \begin{cases} 0 & , \quad k < n \\ (A_{k-n}^{\alpha}/A_k^{\alpha})P_n f & , \quad k \geq n \end{cases}$$

one obtains by Lemma 3.2

$$P_n f^{\eta} = P_n f \sum_{k=n}^{\infty} A_k^{\alpha}(A_{k-n}^{\alpha}/A_k^{\alpha})\Delta^{\alpha+1}\eta_k + \eta_{\infty} P_n f$$

$$= P_n f \{\sum_{k=o}^{\infty} A_k^{\alpha}\ \Delta^{\alpha+1}\eta_{k+n} + \eta_{\infty}\} = \eta_n P_n f .$$

Remark. 1. Muckenhoupt [78] has proved a type of "(C,0)-boundedness" for expansions in Laguerre and Hermite polynomials which may lead to the following modification: As usual, let $\{P_k\}$ decompose X. For some Banach subspace $Y \subset X$ we call η a multiplier from Y into X if to each $f \in Y$ there is associated $f^{\eta} \in X$ such that $P_k f^{\eta} = \eta_k P_k f$. The assumption (3.7) may then be weakened to

$$\text{(3.12)} \qquad \| (C,\alpha)_n f\| \leq C_{\alpha}\| f\|_Y \qquad (n \in P\ ; f \in Y).$$

Indeed, by an analogous argument as in Theorem 3.3 the elements of $bv_{\alpha+1}$ are multipliers from Y into X; but we don't pursue this aspect any further.

2. One may weaken (3.7) in another direction in as far as a certain prescribed growth upon $\| (C,\alpha)_n\|_{[X]}$ is admitted, e.g.

$$\| (C,\alpha)_n f\| \leq \chi_{\alpha}(n)\| f\| \qquad (f \in X)$$

with $\lim_{n\to\infty} \chi_{\alpha}(n) = \infty$. A well-known example in case $\alpha = 0$ is the 1-dimensional Fourier series on $L_{2\pi}^1$ (see Sec. 5.1) for which

$$\| \sum_{k=0}^{n} P_k f \| \leq M \log(n+2) \| f \| \qquad (n \in P)$$

In this case

(3.13) $$\eta \in l^\infty, \ \sum_{k=0}^{\infty} \log(k+2) |\Delta \eta_k| < \infty$$

is a sufficient criterium for $\eta \in M$ (which Moore [75] attributes to Sidon). However, this criterium is disadvantageous in case of a multiplier family which depends multiplicatively upon a parameter, since some type of homogeneity (as indicated in (3.2)) is lacking.

Returning to Theorem 3.3, this criterium may be used to give a growth estimate of $\|P_n\|$ (under hypothesis (3.7)). Indeed, P_n corresponds to the multiplier sequence $\{\delta_{n,k}\}_{k=0}^{\infty}$. Obviously $\Delta^{\alpha+1}\delta_{n,k} = 0$ for $k > n$, and hence

$$\|P_n\| \leq C_\alpha \sum_{m=0}^{n} A_m^\alpha \, |\sum_{k=0}^{n} A_k^{-\alpha-2} \delta_{n,m+k}| = C_\alpha \sum_{m=0}^{n} A_m^\alpha |A_{n-m}^{-\alpha-2}| .$$

If $\alpha = 0$, the latter sum is clearly uniformly bounded in n. For $\alpha > 0$ choose $k \in P$ such that $k < \alpha \leq k+1$. Then $A_{k+2+m}^{-\alpha-2}$ has constant sign for $m \in P$ and $A_m^{-\alpha-2}$ has sign opposite to $A_{m+1}^{-\alpha-2}$, $0 \leq m \leq k+1$. Hence

$$\|P_n\| \leq |\sum_{m=0}^{n} A_m^\alpha \, A_{n-m}^{-\alpha-2}| + 2|\sum_{m=0}^{[(k+1)/2]} A_{n-(k+1+2m)}^\alpha A_{k+1-2m}^{-\alpha-2}| .$$

The first sum on the right hand side is zero for $n \in N$ by (3.5) (and clearly bounded for $n = 0$); the single members of the second sum can be estimated by (3.6) by

$$|A_{k+1-2m}^{-\alpha-2}| \leq 2 \max\{1, \alpha^{k+1}/k!\}, \quad A_{n-(k+1+2m)}^\alpha = O(n^\alpha),$$

and hence under the hypothesis (3.7)

(3.14) $$\|P_n\| \leq D_\alpha (1+n)^\alpha \qquad (n \in P).$$

This simple example already illustrates that a criterium, easier to verify than Theorem 3.3, is highly desirable for concrete summation processes. This is the aim of the next two sections.

3.2 Estimates by integrals in the integer case

As already mentioned in the introduction the main ideas of Sec. 3.2 and 3.3 consist in i) a suitable extension of η (as defined on P) to a function $e(x)$ on the whole half-axis $x \geq 0$ in order to estimate $\|\eta\|_{bv_{\alpha+1}}$ by an integral over some derivatives of $e(x)$ (cf. Butzer-Nessel-Trebels [33;II] for $\alpha = j \in P$), and ii) replacing the variable x in $e(x)$ by a suitable monotone function $\Phi(x)$. To present the approach as lucid as possible we here restrict ourselves to the integer case and, to avoid confusion, we write $\alpha = j \in P$. The fractional case is then treated in the next subsection.

Analogously to [33;II] we introduce the classes

$$(3.15) \qquad BV_{j+1} = \{e \in L_0^\infty; e,\dots,e^{(j-1)} \in AC_{loc}(0,\infty),\ e^{(j)} \in BV_{loc}(0,\infty),$$

$$\|e\|_{BV_{j+1}} = \sup_{x\geq 0} |e(x)| + \frac{1}{\Gamma(j+1)} \int_0^\infty x^j |de^{(j)}(x)| < \infty\},$$

where L_0^∞ is the set [4] of all bounded functions on $[0,\infty)$ with $\lim_{x\to\infty} e(x) = 0$; the latter condition implies no loss of generality, since it is always sufficient to consider $\{\eta_k - \eta_\infty\}_{k=0}^\infty$. $AC_{loc}(0,\infty)$ is the set of locally absolutely continuous functions on $(0,\infty)$(excepting the origin), $BV_{loc}(0,\infty)$ the set of functions, locally of bounded variation on $(0,\infty)$ (excepting the origin), and $L^1_{loc}(0,\infty)$ the set of functions integrable on each compact subset of $(0,\infty)$. Let us point out the fact that consequently $e \in BV_{j+1}$ does not imply the absolute continuity of $e^{(k)}(x)$, $0 \leq k \leq j-1$, at the origin, nor the integrability of $e^{(k)}(x)$, $1 \leq k \leq j$, at the origin. In many cases of interest $e^{(j)} \in AC_{loc}(0,\infty)$; then $\int_0^\infty x^j |de^{(j)}(x)| < \infty$ is satisfied if $\int_0^\infty x^j |e^{(j+1)}(x)|dx < \infty$.

Lemma 3.4. $\qquad BV_{j+1} \subset BV_{k+1} \qquad (0 \leq k \leq j)$.

Proof. Since $e \in BV_{j+1}$, the difference

$$e^{(j)}(y) - e^{(j)}(x) = \int_x^y de^{(j)}(u)$$

tends to zero if $x,y \to \infty$. Thus $\lim_{y\to\infty} e^{(j)}(y) = 1$ exists, and hence

4) The set L_0^∞ is only introduced to include the case $j = 0$.

$$(3.16) \qquad l - e^{(j)}(x) = \int_x^\infty de^{(j)}(u) .$$

If $l = 0$ it follows that

$$\int_0^\infty x^{j-1}|e^{(j)}(x)|\,dx \leq \int_0^\infty x^{j-1} \int_x^\infty |de^{(j)}(u)|\,dx = \int_0^\infty u^j\,|de^{(j)}(u)| ,$$

where the interchange of the integration order is justified by Fubini's theorem on account of the absolute convergence of the double integral. Thus $BV_{j+1} \subset BV_j$.

In order to show $l = 0$, integrate (3.16) (cf. [21]):

$$\int_1^x (x-t)^{j-1} l\,dt - \int_1^x (x-t)^{j-1} e^{(j)}(t)dt = \int_1^x (x-t)^{j-1} \int_t^\infty de^{(j)}(u)dt .$$

An evaluation of the single terms and multiplication by j/x^j gives

$$\frac{(x-1)^j}{x^j}\, l - \frac{j!}{x^j}\, e(x) + \sum_{k=o}^{j-1} \frac{j!}{(j-1-k)!} \frac{(x-1)^{j-1-k}}{x^j}\, e^{(j-1-k)}(1)$$

$$= \int_1^x \{(1-\tfrac{1}{x})^j - (1-\tfrac{u}{x})^j\}\, de^{(j)}(u) + \frac{(x-1)^j}{x^j} \int_x^\infty de^{(j)}(u) .$$

For $x \to \infty$ the right hand side tends to zero by Lebesgue's theorem on dominated convergence and the hypothesis $e \in BV_{j+1}$. Thus, since e is uniformly bounded,

$$l = \lim_{x\to\infty} j!\, x^{-j}\, e(x) = 0 .$$

An appropriate iteration of the above arguments yields the assertion.

<u>*Lemma 3.5.*</u> *Let $e \in BV_{j+1}$ and set*

$$I(x) = \{(-1)^{j+1}/\Gamma(j+1)\} \int_x^\infty (t-x)^j de^{(j)}(t) .$$

Then $e(x) = I(x)$ for all $x \geq 0$.

Proof. Partial integration of $I(x)$ gives for $j \in \mathbb{N}$ and $x > 0$

$$(3.17) \qquad I(x) = (-1)^{j+1} \frac{(t-x)^j}{\Gamma(j+1)} e^{(j)}(t)\Big|_{t=x}^{\infty} + \frac{(-1)^j}{\Gamma(j)} \int_x^{\infty} (t-x)^{j-1} e^{(j)}(t)dt.$$

Now, by the preceding lemma and proof one has

$$|u^j e^{(j)}(u)| \leq u^j \int_u^{\infty} |de^{(j)}(y)| \leq \int_u^{\infty} y^j |de^{(j)}(y)| = o(1)$$

for $u \to \infty$, since $e \in BV_{j+1}$. Thus, the first term on the right side of (3.17) vanishes for each $x > 0$. Iterating $(j-1)$-times gives

$$I(x) = -\int_x^{\infty} e'(t)dt = e(x)$$

for all $x \geq 0$ since the integral is continuous. The assertion for $j = 0$ is trivial.

Lemma 3.6. *If for $\eta \in s$ there exists a function $e \in BV_{j+1}$ such that $\eta_k = e(k)$, $k \in P$, then $\eta \in bv_{j+1}$ and*

$$(3.18) \qquad \sum_{k=0}^{\infty} A_k^j |\Delta^{j+1}\eta_k| \leq (1/\Gamma(j+1)) \int_0^{\infty} x^j |de^{(j)}(x)| ,$$

the constant $1/j!$ being best possible.

Proof. By the preceding lemma

$$\begin{aligned}(-1)^{j+1}\Delta^{j+1}\eta_k &= (1/j!)\sum_{m=0}^{j+1} A_m^{-j-2} \int_{k+m}^{\infty} (t-(k+m))^j de^{(j)}(t) \\ &= (1/j!)\sum_{n=k}^{\infty} \int_n^{n+1} \{\sum_{m=0}^{n-k} A_m^{-j-2}(t-(k+m))^j\} de^{(j)}(t).\end{aligned}$$

Under the assumption that the inner sum is non-negative it follows after an interchange of summation

$$\begin{aligned}\sum_{k=0}^{\infty} A_k^j |\Delta^{j+1}\eta_k| &\leq (1/j!)\sum_{n=0}^{\infty} \int_n^{n+1} \sum_{k=0}^{n} A_k^j \sum_{m=0}^{n-k} A_m^{-j-2}(t-(k+m))^j |de^{(j)}(t)| \\ &\leq (1/j!) \int_0^{\infty} t^j |de^{(j)}(t)|,\end{aligned}$$

for on account of (3.5) one has

$$\sum_{k=0}^{n} A_k^j \sum_{m=0}^{n-k} A_m^{-j-2}(t-(k+m))^j = \sum_{l=0}^{n} (t-l)^j \sum_{k=0}^{l} A_k^j A_{l-k}^{-j-2} = t^j.$$

Thus it remains to prove that

$$(3.19)\qquad \sum_{m=0}^{n-k} A_m^{-j-2}(t-(k+m))^j \geq 0 \qquad (n \leq t \leq n+1, j \in P, 0 \leq k \leq n),$$

which will be achieved by induction. Since $A_0^{-2} = 1$, $A_1^{-2} = -1$, $A_m^{-2} = 0$ for $m \geq 2$, there obviously holds in case $j = 0$

$$\sum_{m=0}^{n-k} A_m^{-2} = \begin{Bmatrix} 1, & k = n \\ 0, & 0 \leq k < n \end{Bmatrix} \geq 0 .$$

Supposing (3.19) to be valid for some j one obtains for $0 \leq k \leq n-1$, $n \leq t \leq n+1$

$$0 \leq (j+1)\{\int_n^t \sum_{m=0}^{n-k} A_m^{-j-2}(u-(k+m))^j du + \int_t^{n+1} \sum_{m=0}^{n-k-1} A_m^{-j-2}(u-(k+1+m))^j du\}$$

$$= \sum_{m=0}^{n-k} A_m^{-j-2}(t-(k+m))^{j+1} - \sum_{m=0}^{n-k-1} A_m^{-j-2}(t-(k+1+m))^{j+1}$$

$$= \sum_{m=0}^{n-k} A_m^{-j-3}(t-(k+m))^{j+1} .$$

Since the statement (3.19) is trivial for $k = n$, (3.19) holds. To show that $1/j!$ is best possible, consider the function $(1+x)^{-1}$ which belongs to BV_{j+1}. Each derivative has constant sign, and omitting the absolute value marks effects a (-1)-factor at most.

Remark. a) A slight modification of this proof immediately shows that

$$(3.20)\qquad \sum_{k=k_0}^{\infty} A_k^j |\Delta^{j+1}\eta_k| \leq 1/j! \int_{k_0}^{\infty} x^j |de^{(j)}(x)| \qquad (k_0 \in P).$$

b) Lemma 3.6 is already proved in [33;II] by another method; it was conjectured there that $1/j!$ is best possible.

It is evident from the general results in Sec. 2 that we usually have to examine families of multipliers. The above lemma may be applied for each fixed $\rho > 0$; but it also gives a uniform multiplier bound if $\{\eta(\rho)\}$ is of Fejér's type, i.e., $\eta_k(\rho) = e(k/\rho)$. Then we have

Theorem 3.7. Let X, $\{P_k\}$ satisfy condition (C^α) (see Def. 3.0) for $\alpha = j \in P$. For $\{\eta(\rho)\} \subset s$ let there exist $e \in BV_{j+1}$ such that $\eta_k(\rho) = e(k/\rho)$. Then $\{\eta(\rho)\}$ is a family of uniformly bounded multipliers and

$$\|\eta(\rho)\|_M \leq C_j \|e\|_{BV_{j+1}} \qquad (\rho > 0) .$$

The condition that $\{\eta(\rho)\}_{\rho>0}$ should be of Fejér's type can be weakened considerably. This together with a further simplification for the applications will provide the following analog of Hardy's "Second theorem of consistency".

Lemma 3.8. *Let $\Psi(\rho)$ be a positive, monotone increasing function on $(0,\infty)$; let $\Phi(t)$ be a non-negative, strongly monotone increasing function with $\lim_{t\to 0+} \Phi(t) = 0$ and $\lim_{t\to\infty} \Phi(t) = \infty$, and let Φ possess $(j+1)$ continuous derivatives on $(0,\infty)$ (the origin is excluded!) with*

$$|t^k \Phi^{(k+1)}(t)| \leq D\, \Phi'(t) \qquad (0 \leq k \leq j).$$

Then, for each $e \in BV_{j+1}$,

$$\int_0^\infty t^j \left| d\left[(\tfrac{d}{dt})^j e(\Phi(t)/\Psi(\rho)) \right] \right| \leq D^* \int_0^\infty x^j |de^{(j)}(x)|,$$

the constant D^* being independent of ρ.

Proof. First observe that

$$d\left[(d/dt)^j\, e(\Phi(t)/\Psi(\rho)) \right]$$

is a finite linear combination (depending on j) of

$$(\Phi'(t)/\Psi(\rho))^j \, de^{(j)}(\Phi(t)/\Psi(\rho))$$

and

$$\{\Psi(\rho)\}^{k-2-j} \prod_{n=1}^{k} \{\Phi^{(n)}(t)\}^{\varepsilon^{n,k}} \, e^{(j+2-k)}(\Phi(t)/\Psi(\rho))dt,$$

where $2 \leq k \leq j+1$, $\varepsilon^{n,k} \in P$ with (cf. [87; p. 20])

$$\sum_{n=1}^{j+1} \varepsilon^{n,k} = j+2-k, \quad \sum_{n=1}^{j+1} n\, \varepsilon^{n,k} = j+1 .$$

Hence, $\int t^j \left| d\left[(d/dt)^j\, e(\Phi(t)/\Psi(\rho))\right]\right|$ may be estimated by a finite linear combination of terms of the type

$$I_1 = \int_0^\infty t^j \{\Phi'(t)/\Psi(\rho)\}^j \, |de^{(j)}(\Phi(t)/\Psi(\rho))| \;,$$

$$I_k = \int_0^\infty t^j \{\Psi(\rho)\}^{k-2-j} \prod_{n=1}^{k} |\Phi^{(n)}(t)|^{\varepsilon^{n,k}} \, |e^{(j+2-k)}(\Phi(t)/\Psi(\rho))| \, dt$$

with $2 \leq k \leq j + 1$. By hypothesis it follows that

$$t\,\Phi'(t) \leq \int_0^t \Phi'(y)dy + \int_0^t y|\Phi''(y)|dy \leq (D+1)\Phi(t),$$

$$(3.21) \qquad t^n|\Phi^{(n)}(t)| \leq Dt\,\Phi'(t) \leq D(D+1)\Phi(t),$$

and therefore

$$I_1 \leq (D+1)^j \int_0^\infty \{\Phi(t)/\Psi(\rho)\}^j \, |de^{(j)}(\Phi(t)/\Psi(\rho))| \leq (D+1)^j \int_0^\infty x^j |de^{(j)}(x)| \;.$$

To estimate I_k choose $n = m$ such that $\varepsilon^{m,k} \neq 0$. Since $\sum_{n=1}^{j+1} n\,\varepsilon^{n,k} = j+1$, $\sum_{n=1}^{j+1} \varepsilon^{n,k} = j + 2 - k$, and by (3.21) it now follows that

$$t^j \prod_{n=1}^{k} |\Phi^{(n)}(t)|^{\varepsilon^{n,k}} \leq t^{m\varepsilon^{m,k}-1} |\Phi^{(m)}(t)|^{\varepsilon^{m,k}} \prod_{n \neq m} |t^n \Phi^{(n)}(t)|^{\varepsilon^{n,k}}$$

$$\leq D_1 \{\Phi(t)\}^{j+1-k} \Phi'(t),$$

and therefore

$$I_k \leq D_1 \int_0^\infty \{\Phi(t)/\Psi(\rho)\}^{j+1-k} |e^{(j+2-k)}(\Phi(t)/\Psi(\rho))| \{\Phi'(t)/\Psi(\rho)\} dt$$

$$= D_1 \int_0^\infty x^{j+1-k} \, |e^{(j+2-k)}(x)| dx \leq D_2 \int_0^\infty x^j |de^{(j)}(x)| \;,$$

the latter estimate following by Lemma 3.4. Thus the proof is complete.

<u>Remark.</u> The proof of the above lemma is a suitable modification of Hirst's [64] proof of the "Second theorem of consistency" for numerical series; here we used some hypotheses on Φ stronger than Hirst which, however, are also quite customary (cf. [39]).

For later use we give some examples of admissible functions Φ:

$$(3.22) \quad \{t(t+\omega)\}^\kappa \; (\omega \geq 0,\ \kappa > 0),\quad t^\alpha \log(1+t) \; (\alpha \geq 0),\quad \log(1+t^\beta) \; (\beta > 0) \;.$$

However, e^t does not satisfy the conditions upon Φ in Lemma 3.8. Terminating the discussion of the integer case we conclude

Theorem 3.9. *Let X, $\{P_k\}$ satisfy condition (C^α) for $\alpha = j \in \mathrm{P}$, and let Φ and Ψ be as in Lemma 3.8. For $\{\eta(\rho)\} \subset s$ let there exist $e \in BV_{j+1}$ such that $\eta_k(\rho) = e(\Phi(k)/\Psi(\rho))$. Then $\{\eta(\rho)\}$ is a family of uniformly bounded multipliers for X, $\{P_k\}$.*

3.3 Estimates by integrals in the fractional case

The main object of this section is to derive analogs of Theorems 3.7 and 3.9 for the fractional case $\alpha > 0$, $\alpha \notin \mathrm{P}$. For this purpose it is necessary to develop a suitable calculus for fractional integration and differentiation. Some of the subsequent results for this calculus are standard in some way or other, but unfortunately most not in the precise form to be needed later on. For this reason and for the sake of completeness Sec. 3.3 will be divided into two subsections, the first giving the theory on fractional integration and differentiation as needed on $BV_{\alpha+1}$-spaces, the second yielding the desired analogs to Theorems 3.7 and 3.9 as well as some examples.

3.3.1 Fractional integration and differentiation on $BV_{\alpha+1}$, $\alpha \notin \mathrm{P}$

First introduce the following fractional integral operator

$$(3.23)\qquad I_\omega^\alpha[d\mu](x) = \frac{1}{\Gamma(\alpha)} \int_x^\omega (t-x)^{\alpha-1} d\mu(t) \qquad (0 < x < \omega),$$

where $0 < \alpha < 1$, $0 < \omega < \infty$ and μ is a Borel measure, bounded on each compact set of $(0,\infty)$; analogously

$$(3.24)\qquad I_\omega^\alpha[e](x) = \frac{1}{\Gamma(\alpha)} \int_x^\omega (t-x)^{\alpha-1} e(t)\,dt \qquad (0 < x < \omega)$$

for locally integrable functions e on $(0,\infty)$ (excepting the origin). Then it is easy to show that $I_\omega^\alpha[d\mu](x)$ and $I_\omega^\alpha[e](x)$ exist as locally integrable functions on $(0,\infty)$. By $J^\alpha[d\mu]$ (or $J^\alpha[e]$) denote the corresponding Weyl integral

$$(3.25)\qquad J^\alpha[d\mu](x) = \lim_{\omega\to\infty} I_\omega^\alpha[d\mu](x) \qquad (= \lim_{\omega\to\infty} I_\omega^\alpha[e](x)),$$

if the right-hand side exists.

With Cossar [41] define for suitable e a fractional derivative of order α, $0 < \alpha < 1$, by

$$e^{(\alpha)}(x) = \lim_{\omega\to\infty} - \frac{d}{dx} I_\omega^{1-\alpha} e(x). \tag{3.26}$$

Now the first result is due to Cossar [41].

Lemma 3.10. *Let $e \in C_0$, i.e., e is uniformly continuous on $[0,\infty)$ with $\lim_{x\to\infty} e(x) = 0$, and let $e^{(\alpha)} \in L^1_{loc}(0,\infty)$. Then*

$$\lim_{\omega\to\infty} I_\omega^\alpha e^{(\alpha)}(x) = e(x)$$

for almost all x in $(0,\infty)$.

Proof (cf. [41]). Since e is bounded, it follows for $0 < t < \omega$ (ω large) by the definition of $e^{(\alpha)}(x)$ that

$$e^{(\alpha)}(t) = \frac{-1}{\Gamma(1-\alpha)} \frac{d}{dt} \int_t^\omega (y-t)^{-\alpha} e(y)dy - \frac{\alpha}{\Gamma(1-\alpha)} \int_\omega^\infty (y-t)^{-\alpha-1} e(y)dy .$$

Using the formula ($\alpha,\beta > 0$)

$$\int_x^y (y-t)^{\alpha-1}(t-x)^{\beta-1}dt = (y-x)^{\alpha+\beta-1} \frac{\Gamma(\alpha)\Gamma(\beta)}{\Gamma(\alpha+\beta)} , \tag{3.27}$$

it follows that

$$\left| I_\omega^\alpha \left[\frac{\alpha}{\Gamma(1-\alpha)} \int_\omega^\infty (y-t)^{-\alpha-1} e(y)dy \right](x) \right|$$

$$\leq \sup_{t\geq\omega} |e(t)| \frac{1}{\Gamma(1-\alpha)\Gamma(\alpha)} \int_x^\omega (\omega-t)^{-\alpha}(t-x)^{\alpha-1}dt = \sup_{t\geq\omega} |e(t)| .$$

By hypothesis, the last term tends to zero, uniformly in x for $\omega \to \infty$. Thus one has to evaluate

$$d_\omega(x) \equiv I_\omega^\alpha \left[- \frac{1}{\Gamma(1-\alpha)} \frac{d}{dt} \int_t^\omega (y-t)^{-\alpha} e(y)dy \right](x) .$$

First integrate $d_\omega(x)$ over (x,ω):

$$\int_x^\omega d_\omega(z)dz = \frac{1}{\Gamma(\alpha+1)} \int_x^\omega (t-x)^\alpha \{- \frac{1}{\Gamma(1-\alpha)} \frac{d}{dt} \int_t^\omega (y-t)^{-\alpha} e(y)dy\} dt$$

$$= - \frac{1}{\Gamma(\alpha+1)} \frac{(t-x)^\alpha}{\Gamma(1-\alpha)} \int_t^\omega (y-t)^{-\alpha} e(y)dy \Big|_{t=x}^{\omega}$$

$$+ \frac{1}{\Gamma(\alpha)} \int_x^\omega (t-x)^{\alpha-1} \frac{1}{\Gamma(1-\alpha)} \int_t^\omega (y-t)^{-\alpha} e(y)dy \; dt \quad .$$

Since e is bounded it follows that $I_\omega^{1-\alpha}[e](\omega) = 0$, and $I_\omega^{1-\alpha}[e]\ (x)$ is bounded for each x, $0 < x < \omega$. Thus the first term on the right-hand side vanishes.

Another interchange of integration order (which is possible on account of the absolute convergence of the double integral) finally yields

$$\int_x^\omega d_\omega(y)dy = \frac{1}{\Gamma(\alpha)\Gamma(1-\alpha)} \int_x^\omega e(y)dy \int_x^y (t-x)^{\alpha-1}(y-t)^{-\alpha}dt = \int_x^\omega e(y)dy \; .$$

Hence, for almost all x in $0 < x < \omega$ (ω fixed) there holds $d_\omega(x) = e(x)$ Denoting by A_ω the exceptional set (where the latter equality is not valid) one has $m\ (A_\omega) = 0$ (m being the Lebesgue measure) and

$$(3.28) \qquad |I_\omega^\alpha[e^{(\alpha)}](x) - e(x)| \leq \sup_{t \geq \omega} |e(t)|$$

for all $x \in \{(0,\omega - 1) \setminus A_\omega\}$. Let Ω be a countable set dense in $(0,\infty)$. Then $\bigcup_{\omega \in \Omega} A_\omega$ is of Lebesgue measure zero. Hence by (3.28) and the hypothesis $e \in C_0$ one has

$$\lim_{\omega\to\infty, \omega\in\Omega} I_\omega^\alpha[e^{(\alpha)}](x) = e(x) \qquad \text{a.e.}$$

Since, however, $I_\omega^\alpha[e^{(\alpha)}](x)$ is a continuous function of ω, it will tend to the same limit when ω is unrestricted. The result now follows.

To avoid unnecessary accessories we restrict the set of functions e to those belonging to $BV_{\alpha+1}$ (the aim still being estimates of $\|\eta\|_{bv_{\alpha+1}}$). First, let us consider the case $0 < \alpha < 1$:

$$(3.29) \quad BV_{\alpha+1} = \{e \in C_0;\ e^{(\alpha)} \in BV_{loc}(0,\infty) \quad \text{such that}$$

$$\|e\|_{BV_{\alpha+1}} = \sup_{x \geq 0} |e(x)| + \frac{1}{\Gamma(\alpha+1)} \int_0^\infty x^\alpha |de^{(\alpha)}(x)| < \infty\} .$$

Lemma 3.11. $e \in BV_{\alpha+1}$, $0 < \alpha < 1$, *implies that for all* $x \in [0,\infty)$

$$e(x) = -\frac{1}{\Gamma(\alpha+1)} \int_x^\infty (y-x)^\alpha \, de^{(\alpha)}(y).$$

Furthermore, $e'(x) = J^\alpha[de^{(\alpha)}](x)$ *a.e. and* $e' \in L^1(0,\infty)$.

Proof. Partial integration yields

$$(3.30) \qquad \frac{1}{\Gamma(\alpha+1)} \int_x^\omega (y-x)^\alpha \, de^{(\alpha)}(y)$$
$$= \frac{(y-x)^\alpha e^{(\alpha)}(y)}{\Gamma(\alpha+1)}\Big|_{y=x}^{\omega} - \frac{1}{\Gamma(\alpha)} \int_x^\omega (y-x)^{\alpha-1} e^{(\alpha)}(y)\,dy .$$

By the same argument as in Lemma 3.4 it follows that

$$e^{(\alpha)}(x) = 1 - \int_x^\infty de^{(\alpha)}(u)$$

and hence, by the hypothesis $e \in BV_{\alpha+1}$, that $e^{(\alpha)}(x)$ is uniformly bounded in $x \geq a$ for each fixed $a > 0$. Thus the first term on the right side of (3.30) vanishes at $y = x$. The second term on the right side of (3.30) tends to $e(x)$ almost everywhere provided $\omega \to \infty$. The left hand side of (3.30) is uniformly bounded. Thus, $\lim_{\omega\to\infty} (\omega-x)^\alpha e^{(\alpha)}(\omega)$ exists and is bounded, hence $\lim_{\omega\to\infty} e^{(\alpha)}(\omega) = 0$. Then, by the same arguments as in Lemma 3.4,

$$\lim_{\omega\to\infty} \omega^\alpha |e^{(\alpha)}(\omega)| = 0 .$$

Now passing to the limit $\omega \to \infty$ in (3.30) the representation of $e(x)$ almost everywhere in $(0,\infty)$ follows by Lemma 3.10.

To prove equality for all $x \in [0,\infty)$ it remains to show the continuity of

$$(3.31) \qquad I(x) \equiv \frac{1}{\Gamma(\alpha+1)} \int_x^\infty (y-x)^\alpha de^{(\alpha)}(y)$$
$$= \frac{1}{\Gamma(\alpha)} \int_x^\infty du \int_u^\infty (y-u)^{\alpha-1} \, de^{(\alpha)}(y) .$$

Now

$$\int_0^\infty \left| \int_u^\infty (y-u)^{\alpha-1} de^{(\alpha)}(y) \right| dy \leq \int_0^\infty |de^{(\alpha)}(y)| \int_0^y (y-u)^{\alpha-1} du$$
$$= \frac{1}{\alpha} \int_0^\infty y^\alpha |de^{(\alpha)}(y)| < \infty ,$$

the interchange of integration being justified by Fubini's theorem on account of the absolute convergence. Thus, the inner integral in (3.31) is integrable on $(0,\infty)$ (including the origin) and hence $I(x)$ is differentiable, a fortiori continuous. Differentiation of $I(x)$ finally yields the rest of the assertion.

Lemma 3.12. *Let $e \in BV_{\alpha+1}$ and $0 < \delta < \alpha \leq 1$. Then $e^{(\delta)}(x)$, $e^{(\delta+1)}(x)$ exist almost everywhere, $e^{(\delta)}$, $e^{(\delta+1)} \in L(a,\infty)$ for each fixed $a > 0$ and*

$$e^{(\delta)}(x) = -J^{1+\alpha-\delta}[\,de^{(\alpha)}](x), \quad e^{(\delta+1)}(x) = J^{\alpha-\delta}[\,de^{(\alpha)}](x) \qquad a.e.$$

Proof. By definition (3.26) of $e^{(\delta)}$ one has to examine $\lim_{\omega\to\infty}(-d/dx)\, I_\omega^{1-\delta}[e](x)$. By Lemmata 3.11 and 3.5 one knows that $e(y) = -\int_y^\infty e'(t)dt$, and therefore

$$I_\omega^{1-\delta}[e](x) = -\frac{1}{\Gamma(1-\delta)} \int_x^\omega (y-x)^{-\delta} \int_y^\omega e'(t)dt\, dy$$

$$- \frac{1}{\Gamma(1-\delta)} \int_x^\omega (y-x)^{-\delta} \int_\omega^\infty e'(t)dt\, dy \equiv I_1(x) + I_2(x)\ .$$

Obviously it follows that

$$-\frac{d}{dx} I_2(x) = \frac{d}{dx} e(\omega) \frac{(\omega-x)^{1-\delta}}{\Gamma(2-\delta)} = -\frac{(\omega-x)^{-\delta}}{\Gamma(1-\delta)} e(\omega)\ .$$

Passing to the limit $\omega \to \infty$, this expression tends to zero uniformly in x, $0 < a \leq x \leq \omega_0 < \omega < \infty$, a and ω_0 fixed. In $I_1(x)$ we interchange the order of integration to obtain

$$I_1(x) = -\int_x^\omega e'(t)dt \frac{1}{\Gamma(1-\delta)} \int_x^t (y-x)^{-\delta}dy$$

$$= -\int_x^\omega \frac{(t-x)^{1-\delta}}{\Gamma(2-\delta)} e'(t)dt = -\int_x^\omega dy \int_y^\omega \frac{(t-y)^{-\delta}}{\Gamma(1-\delta)} e'(t)dt\ .$$

Since the inner integral is locally integrable with respect to y it follows that

$$-(d/dx)I_1(x) = -\frac{1}{\Gamma(1-\delta)} \int_x^\omega (t-x)^{-\delta} e'(t)dt \qquad a.e.$$

Now replace e' by the representation of Lemma 3.11 to deduce for $\omega \to \infty$

$$e^{(\delta)}(x) = \lim_{\omega\to\infty} - \int_x^\omega \frac{(t-x)^{-\delta}}{\Gamma(1-\delta)} \frac{1}{\Gamma(\alpha)} \int_t^\infty (y-t)^{\alpha-1} de^{(\alpha)}(y) \qquad \text{a.e.}$$

Since

$$\frac{1}{\Gamma(1-\delta)} \int_x^\infty (t-x)^{-\delta} \frac{1}{\Gamma(\alpha)} \int_t^\infty (y-t)^{\alpha-1} |de^{(\alpha)}(y)| =$$

$$= \frac{1}{\Gamma(1+\alpha-\delta)} \int_x^\infty (y-x)^{\alpha-\delta} |de^{(\alpha)}(y)|$$

is bounded for $x \geq a > 0$ by hypothesis, the former limit exists for each $x > 0$, defines an absolutely convergent improper integral, and thus a continuous function which we identify with $e^{(\delta)}$ by definition (the extension of the domain $a \leq x \leq \omega_o$ to $0 < x < \infty$ may be carried out as in Lemma 3.11). Finally, since

$$e^{(\delta)}(x) = -J^{1+\alpha-\delta}[de^{(\alpha)}](x) = -\frac{1}{\Gamma(\alpha-\delta)} \int_x^\infty dt \int_t^\infty (y-t)^{\alpha-\delta-1} de^{(\alpha)}(y),$$

and the inner integral on the right hand side is integrable over $[a,\infty)$ for each fixed $a > 0$, a further differentiation of $e^{(\delta)}(x)$ gives the rest of the assertion.

Now we can easily derive an extension of Lemma 3.4 for $j = 1$ to the fractional case, namely

Corollary 3.13. $\quad BV_{\alpha+1} \subset BV_{\delta+1} \qquad (0 \leq \delta < \alpha \leq 1).$

Proof. By Lemmata 3.4 - 5, 3.11 - 12, $e \in BV_{\alpha+1}$ implies the existence of $e^{(\delta+1)} \in L(a,\infty)$, $a > 0$, and thus

$$\frac{1}{\Gamma(\delta+1)} \int_a^\infty x^\delta |e^{(\delta+1)}(x)| dx \leq \frac{1}{\Gamma(\delta+1)} \int_a^\infty x^\delta \frac{1}{\Gamma(\alpha-\delta)} \int_x^\infty (y-x)^{\alpha-\delta-1} |de^{(\alpha)}(y)|$$

$$\leq \int_a^\infty |de^{(\alpha)}(y)| \frac{1}{\Gamma(\delta+1)} \frac{1}{\Gamma(\alpha-\delta)} \int_o^y x^\delta (y-x)^{\alpha-\delta-1} dx = \frac{1}{\Gamma(\alpha+1)} \int_a^\infty y^\alpha |de^{(\alpha)}(y)| .$$

Passing to the limit for $a \to 0+$ yields the assertion.

As usual (cf. [110;II,p.134]), fractional derivatives of higher order $\alpha = \gamma + n$, $0 < \gamma < 1$ and $n \in \mathbb{N}$, are defined by

$$e^{(\alpha)}(x) = (\tfrac{d}{dx})^n e^{(\gamma)}(x) \tag{3.32}$$

for sufficiently smooth functions e. Analogously to (3.29) we introduce for $\alpha > 1$ the set

$$(3.33)\quad BV_{\alpha+1} = \{ e \in C_o;\ e^{(\gamma)},\dots,e^{(\alpha-1)} \in AC_{loc}(0,\infty),\ e^{(\alpha)} \in BV_{loc}(0,\infty),$$

$$\text{and } \|e\|_{BV_{\alpha+1}} = \sup_{x\geq o}|e(x)| + \frac{1}{\Gamma(\alpha+1)} \int_o^\infty x^\alpha |de^{(\alpha)}(x)| < \infty \} .$$

With the aid of the above results in the case $0 < \alpha \leq 1$ it is not difficult to prove

Lemma 3.14. $BV_{\alpha+1} \subset BV_{\alpha+1-k}$ $\quad (\alpha > 1,\ \alpha \notin \mathbb{N},\ 1 \leq k \leq [\alpha])$, *and each $e \in BV_{\alpha+1}$ can be represented by*

$$e(x) = \frac{(-1)^{[\alpha]+1}}{\Gamma(\alpha+1)} \int_x^\infty (t-x)^\alpha de^{(\alpha)}(t) .$$

Proof. As in the proof of Lemma 3.4 one has

$$e^{(\alpha)}(t) = l - \int_t^\infty de^{(\alpha)}(y) ,$$

and hence $BV_{\alpha+1} \subset BV_\alpha$ provided $l = 0$. Assume $l \neq 0$, and without loss of generality that $l > 0$. Then there exists $t_o > 0$ such that $e^{(\alpha)}(t) \geq l/2$ for all $t \geq t_o$; $[\alpha]$-fold integration over $(1,t)$ yields

$$e^{(\alpha-[\alpha])}(t) \geq \frac{l}{2}\,\frac{(t-1)^{[\alpha]}}{[\alpha]!} - O\big((t-1)^{[\alpha]-1}\big) \geq (l/4[\alpha]!)(t-1)^{[\alpha]}$$

for sufficiently large t. Since $e^{(\alpha-[\alpha])}$ is locally integrable, one obtains by Lemma 3.10

$$e(x) = \lim_{\omega\to\infty} I_\omega^{\alpha-[\alpha]}[e^{(\alpha-[\alpha])}](x)$$

$$\geq (l/4[\alpha]!)(x-1)^{[\alpha]} \lim_{\omega\to\infty} \frac{1}{\Gamma(\alpha-[\alpha])} \int_x^\omega (t-x)^{\alpha-[\alpha]-1} dt$$

which obviously tends to infinity with $x \to \infty$, in contradiction to the hypothesis $e \in C_o$. Thus $l = 0$ and iteration repeated $([\alpha]-1)$-times gives the first part of the assertion.

Hence we may integrate

$$I(x) = \frac{1}{\Gamma(\alpha+1)} \int_x^\infty (t-x)^\alpha de^{(\alpha)}(t)$$

by parts to obtain by the same procedure as in the proof of Lemma 3.11 that

$$I(x) = \frac{-1}{\Gamma(\alpha)} \int_x^\infty (t-x)^{\alpha-1} e^{(\alpha)}(t)dt .$$

An $([\alpha]-1)$-fold iteration of these arguments leads to

$$I(x) = \frac{(-1)^{[\alpha]}}{\Gamma(1+\alpha-[\alpha])} \int_x^\infty (t-x)^{\alpha-[\alpha]} e^{(\alpha-[\alpha]+1)}(t)dt = (-1)^{[\alpha]+1} e(x)$$

by Lemma 3.11.

To round off a little the results on fractional differentiation over BV-spaces we finally prove

Lemma 3.15. *i) If* $e \in BV_{\alpha+1}$, *then* $e^{(\gamma)}(x) \in L^1_{loc}(0,\infty)$, $0 < \gamma < \alpha + 1$, *and*

$$e^{(\gamma)}(x) = \pm J^{\alpha-\gamma} [de^{(\alpha)}](x) \qquad a.e.;$$

ii)

$$BV_{\alpha+1} \subset BV_{\beta+1} \qquad (0 \leqslant \beta < \alpha);$$

iii) if $e \in BV_{\alpha+1}$ *and* $e^{(\alpha)} \in AC_{loc}(0,\infty)$, $\beta = n + \delta$ *with* $1 < \beta \leqslant \alpha + 1$, $0 < \delta < 1$ *and* $\alpha > 0$, *then*

$$(e^{(\delta)})^{(n)}(x) = (e^{(n)})^{(\delta)}(x) \qquad a.e.$$

Proof. i) First assume $k < \gamma < k + 1$ and $n < \alpha \leqslant n + 1$, $k, n \in P$. Then, by Lemma 3.12, $e^{(\gamma-k)}$ exists and can be represented by

$$e^{(\gamma-k)}(x) = -J^{1+\alpha-n-\gamma+k}[de^{(\alpha-n)}](x) = (-1)^{n+1} J^{1+\alpha-\gamma+k}[de^{(\alpha)}](x)$$

by n-fold partial integration. Since $\alpha + 1 - \gamma > 0$, k-fold differentiation is possible, and the same arguments as in the proof of Lemma 3.12 yield i) provided $\gamma \notin P$. If $\gamma \in N$, apply Lemmata 3.5 and 3.11, and proceed analogously.

ii) follows immediately by i) since

$$\frac{1}{\Gamma(\beta+1)} \int_0^\infty x^\beta |e^{(\beta+1)}(x)| dx \leqslant \frac{1}{\Gamma(\beta+1)} \int_0^\infty x^\beta J^{\alpha-\beta}[|de^{(\alpha)}|](x)dx$$

$$= \int_0^\infty |de^{(\alpha)}(t)| \int_0^t \frac{x^\beta}{\Gamma(\beta+1)} \frac{(t-x)^{\alpha-\beta-1}}{\Gamma(\alpha-\beta)} dx = \frac{1}{\Gamma(\alpha+1)} \int_0^\infty t^\alpha |de^{(\alpha)}(t)| ,$$

the interchange of integration being justified by Fubini.

iii) By i) the derivative $e^{(n)}(x)$ exists almost everywhere, and by ii) and Lemmata 3.5 and 3.11 one has (after an (n-1)-fold integration by parts and (n-1)-fold differentiation just as in i) above - only the signs cancel each other)

$$e^{(n)}(x) = J^\delta[e^{(\beta)}](x) \qquad \text{a.e.}$$

By definition (3.26) one has to show that

$$(e^{(n)})^{(\delta)}(x) = -\lim_{\omega\to\infty} \frac{d}{dx} I_\omega^{1-\delta}[e^{(n)}](x)$$

exists and to compute its value. Using the above representation of $e^{(n)}$ one obtains

$$I_\omega^{1-\delta}[J^\delta[e^{(\beta)}]](x) = \int_x^\infty e^{(\beta)}(t)dt$$
$$- \int_\omega^\infty e^{(\beta)}(y) \int_\omega^y \frac{(t-x)^{-\delta}}{\Gamma(1-\delta)} \frac{(y-t)^{\delta-1}}{\Gamma(\delta)} dt\, dy = I_1(x) + I_2(x).$$

Obviously, $\lim_{\omega\to\infty}(d/dx)I_1(x) = -e^{(\beta)}(x)$, thus one has to consider $I_2(x)$, where - analogously to Lemma 3.10 - one may suppose that $x < \omega - 1$, ω large. Then it is possible to differentiate $I_2(x)$ (under the integral signs); passing to the limit for $\omega \to \infty$ yields

$$\lim_{\omega\to\infty}|(d/dx)I_2(x)| \leq \lim_{\omega\to\infty} \frac{(\omega-x)^{-\delta-1}}{|\Gamma(-\delta)|} \int_\omega^\infty \frac{(y-\omega)^\delta}{\Gamma(\delta+1)} |e^{(\beta)}(y)|dy = 0$$

(uniformly in x since the latter integral vanishes). Thus the proof is complete.

3.3.2 Sufficient multiplier criteria for fractional α

The analogs of Lemmata 3.4 and 3.5 now being available, one would expect that the same technique as in Lemma 3.6 works for $\alpha \notin P$. This is true with one restriction: unfortunately the author did not succeed in showing whether a result of type (3.19) is valid or not; this drawback requires a modification which is supplied by Gergen's [50] proof on the equivalence of Cesàro's and Riesz' summability methods for numerical series, as already mentioned in the introduction. For the sake of simplicity we restrict ourselves to the pure fractional case $\alpha > 0$, $\alpha \notin P$.

<u>*Lemma 3.16.*</u> *Let $\alpha > 0$ and $\alpha \notin P$. Let $\eta \in s$ be a sequence for which there exists a function $e \in BV_{\alpha+1}$ such that $\eta_k = e(k)$ for all $k \in P$. Then $\eta \in bv_{\alpha+1}$ and*

$$\sum_{k=0}^{\infty} A_k^{\alpha} \, |\Delta^{\alpha+1} \eta_k| \leq C_\alpha^* \int_0^\infty x^\alpha |de^{(\alpha)}(x)|$$

the constant [5)] *C_α^* being independent of η.*

Proof. Applying Lemma 3.14 and proceeding as in the proof of Lemma 3.6 one has

$$\Gamma(\alpha+1)|\Delta^{\alpha+1}\eta_k| \leq \sum_{n=k}^{\infty} \int_n^{n+1} |\sum_{m=0}^{n-k} A_m^{-\alpha-2}(t-(k+m))^\alpha| \, |de^{(\alpha)}(t)|.$$

Multiplying by A_k^α, summing over all k and interchanging the summation order yields

$$\Gamma(\alpha+1)\sum_{k=0}^{\infty} A_k^\alpha|\Delta^{\alpha+1}\eta_k|$$

$$\leq \sum_{n=0}^{\infty} \int_n^{n+1} \sum_{k=0}^{n} A_k^\alpha |\sum_{m=0}^{n-k} A_m^{-\alpha-2}(t-(k+m))^\alpha| \, |de^{(\alpha)}(t)|$$

$$\leq D_1 \sum_{n=0}^{\infty} \int_n^{n+1} t^\alpha (1+\sum_{k=0}^{n-1} | \sum_{m=0}^{n-k} A_m^{-\alpha-2}(t-(k+m))^\alpha| \, |de^{(\alpha)}(t)|$$

by (3.6), noting that $A_k^\alpha \leq A_n^\alpha$, $0 \leq k \leq n$. Hence there remains to estimate

$$\Sigma_n \equiv \sum_{k=0}^{n-1} |\sum_{m=0}^{n-k} A_m^{-\alpha-2}(t-(k+m))^\alpha| \tag{3.34}$$

uniformly in n and $n \leq t \leq n + 1$. This has been essentially achieved by Gergen [50], whose proof we reproduce briefly.

---

5) The integer case induces one to conjecture that $1/\Gamma(\alpha+1)$ is the best constant.

Consider the particular function

$$b_o^*(x) = A_i^{-\alpha-1}\,\Gamma(\alpha+1) ,$$

where i is the largest integer less than x (with x large). Then, for each n (less than x), after an n-fold integration

$$b_n^*(x) \,/\, \Gamma(\alpha+1) = A_{i-n}^{-\alpha-1+n} + (1/n!)\sum_{m=o}^{n-1} A_{i-m}^{-\alpha-1+m}\,\Delta_m(x-i+m)^n,$$

where

$$\Delta_m x^n = \sum_{j=o}^{m} A_j^{-m-1}(x-j)^n$$

is a "backward" difference (cf. in contrast (3.10)). By (3.6) one thus arrives at

$$b_n^*(x)/\Gamma(\alpha+1) = x^{n-\alpha-1}/\Gamma(n-\alpha) + O(x^{n-\alpha-2} + \sum_{m=o}^{n-1} x^{-\alpha-1+m})$$

$$= x^{n-\alpha-1}/\Gamma(n-\alpha) + O(x^{n-\alpha-2}) .$$

Taking

$$b_o(x) = b^*_{[\alpha]}(x)/\Gamma(\alpha+1) \qquad (\alpha = [\alpha] + \delta),$$

one easily verifies

$$b_\delta(x) = b_\alpha^*(x)/\Gamma(\alpha+1) = \frac{1}{\Gamma(\alpha+1)}\sum_{m<x} A_m^{-\alpha-2}(x-m)^\alpha$$

on account of the addition formula for fractional Riemann-Liouville integration (cf.e.g. [85; p. 10]). Furthermore,

$$b_l(x) = x^{l-\delta-1}/\Gamma(l-\delta) + O(x^{l-\delta-2})$$

for $l = 0, 1, 2$, so that Lemma 3.17 (to follow) yields $b_\delta(x) = O(x^{-2})$ for large x. Returning to (3.34) it is sufficient to consider $\sum_n$ for large n; thus

$$\sum_n = O(\sum_{k=o}^{n-1} (n-k)^{-2}) = O(\sum_{k=1}^{n} k^{-2}) = O(1)$$

uniformly in n, and the proof of Lemma 3.16 is complete up to Gergen's [50] lemma (applied above) which we quote without proof.

Lemma 3.17. *Let $0 < \delta < 1$. Let $b_o(x)$ be integrable over every finite interval $0 < x < x_o$ and satisfy*

$$b_l(x) = x^{l-\delta-1}/\Gamma(l-\delta) + O(x^{l-\delta-2}) \qquad (x \to \infty)$$

for $l = 0,1,2$, where

$$b_\beta(x) = \frac{1}{\Gamma(\beta)} \int_0^x (x-t)^{\beta-1}\, b_o(t)dt \qquad (\beta > 0)$$

is the standard Riemann-Liouville fractional integral. Then

$$b_\delta(x) = O(x^{-2}) \qquad (x \to \infty) \ .$$

With the aid of Lemma 3.16 it is now easy to show that the converses of Lemmata 3.2 and 3.15,ii) are not true. This is attained by an example due to Moore [75], which further illustrates that the integral estimate in Lemma 3.16 is quite sharp (if the extension of η to e is "well" choosen). Consider

$$\eta \in s, \ \eta_k = \sum_{m=[1+k/2]} m^{-2}$$

and its extension to the half-axis $x \geq 0$

$$e(x) = \begin{cases} \sum_{n=m}^{\infty} n^{-2} \ , & 2(m-1) \leq x \leq 2m-1 \\ \sum_{n=m}^{\infty} n^{-2} - m^{-2}(x-2m-1), & 2m-1 \leq x \leq 2m \end{cases} \qquad m \in \mathbb{N}$$

It is not hard to show that $e \in BV_{\alpha+1}$ for $0 \leq \alpha < 1$. By Lemma 3.16 we conclude that $\eta \in bv_{\alpha+1}$, and an elementary calculation finally gives $\eta \notin bv_2$ and thus $e \notin BV_2$.

Incited by the general theory in Sec. 2, our main interest naturally lies in uniform estimates for families of sequences. Observing that for each $e \in BV_{\alpha+1}$, $\alpha > 0$, $\rho > 0$

$$(\tfrac{d}{dx})^\alpha e(x/\rho) = \rho^{-\alpha}\, e^{(\alpha)}(x/\rho),$$

this problem is immediately settled for sequences of Fejér's type and, by Lemma 3.16 and Theorem 3.7, we arrive at the following generalization of Theorem 3.7.

Theorem 3.18. *Let X, $\{P_k\}$ satisfy condition (C^{α}) (see Def. 3.0) for some $\alpha \geq 0$. Let $\{\eta(\rho)\}_{\rho>0} \subset s$ be a family of sequences for which there exists a function $e(x) \in BV_{\alpha+1}$ such that $\eta_k(\rho) = e(k/\rho)$ for all $k \in \mathbb{P}$, $\rho > 0$. Then $\{\eta(\rho)\}$ is a family of uniformly bounded multipliers and*

$$\|\eta(\rho)\|_M \leq C_{\alpha}^{*} \|e\|_{BV_{\alpha+1}} \qquad (\rho > 0).$$

As an immediate consequence of this theorem we show that the Riesz means $\mathcal{R}_{1,\beta}$, defined by

$$R_{1,\beta}(\rho)f = \sum_{k<\rho} \left(1-\frac{k}{\rho}\right)^{\beta} P_k f , \tag{3.35}$$

are uniformly bounded in ρ for $\beta \geq \alpha$ provided (3.7) holds for α. Furthermore, by the Banach-Steinhaus theorem, $\mathcal{R}_{1,\beta}$ is a strong approximation process on X. On account of Theorem 3.18 we only have to prove that

$$e_{\beta}(x) = \begin{cases} (1-x)^{\beta}, & 0 \leq x \leq 1 \\ 0, & x \geq 1 \end{cases}$$

belongs to $BV_{\alpha+1}$. Now simple calculations show

$$e_{\beta}^{(\alpha)}(x) = (-1)^{[\alpha]+1} \begin{cases} \Gamma(\beta+1) \left\{ \begin{array}{ll} 1, & \beta = \alpha \\ \dfrac{(1-x)^{\beta-\alpha}}{\Gamma(1+\beta-\alpha)}, & \beta > \alpha \end{array} \right\}, & 0 < x < 1 \\ 0, & x > 1 \end{cases}$$

and thus $e \in BV_{\alpha+1}$.

Theorem 3.19. *Let X and $\{P_k\}$ be as in Sec. 2. Then for $\alpha \geq 0$*

$$\|(C,\alpha)_n f\| \leq C_{\alpha} \|f\| \qquad (f \in X) \tag{3.36}$$

if and only if

$$\|R_{1,\alpha}(\rho) f\| \leq C_{\alpha}^{*} \|f\| \qquad (f \in X). \tag{3.37}$$

Proof. By the preceding it remains to prove that (3.37) implies (3.36). This again has been accomplished essentially by Gergen [50] (for numerical series), whose proof we reproduce. Consider a function e(x) continuous with its derivatives $e',\dots,e^{([\alpha]+2)}$ for $x \geq 0$ and satisfying

$$(3.38) \qquad e(0) = e'(0) = \dots = e^{([\alpha]+1)}(0) = 0,$$

$$(3.39) \qquad e(x) = \Gamma(\alpha+x)/\{\Gamma(x)\Gamma^2(\alpha+1)\} \qquad \text{for } x \geq 1.$$

Define the function

$$u(x) = 1/\Gamma([\alpha]+1-\alpha) \int_0^x (x-t)^{[\alpha]-\alpha} e^{([\alpha]+2)}(t)dt$$

(which actually is a Riemann-Liouville derivative of order $\alpha + 1$; cf. the contrast to the Weyl derivative (3.26), (3.32)). Then, analogously to Lemmata 3.5, 3.11, 3.14, one has (cf. [101])

$$e(x) = 1/\Gamma(\alpha+1) \int_0^x (x-t)^\alpha u(t)dt$$

(here the initial conditions (3.38) are used). Thus it follows that for the Bochner integral

$$\int_0^{n+1} \sum_{k<t}(t-k)^\alpha P_k f \; u(n+1-t)dt$$
$$= \sum_{k=0}^n P_k f \int_0^{n+1-k} (n+1-k-t)^\alpha \, u(t)dt = \sum_{k=0}^n A_{n-k}^\alpha \, P_k f$$

and hence, by (3.37),

$$\|(C,\alpha)_n f\| \leq \int_0^{n+1} (t^\alpha/A_n^\alpha)\| R_{1,\alpha}(t)f\| \, |u(n+1-t)|dt$$
$$\leq D_\alpha \|f\| \int_0^\infty |u(t)|dt.$$

Hence it remains to establish the boundedness of the latter integral. Since u is continuous, only its behaviour at infinity is of interest. This again will be settled by an application of Lemma 3.17.

If α is an integer, then e reduces to a polynomial of degree α for $x \geq 1$. Hence

$$u(x) = e^{(\alpha+1)}(x) = 0 \qquad (x > 1),$$

and the integral in question converges.

Suppose then that α is not an integer and set $[\alpha]+1-\alpha = \delta$, $0 < \delta < 1$. From Stirling's formula (see [17; pp.297-308])

$$\log\Gamma(x) = (x - \tfrac{1}{2})\log x - x + \tfrac{1}{2}\log 2\pi + \frac{P(0)}{x} - \int_0^\infty \frac{P(t)}{(t+x)^2}\,dt ,$$

$$P(t) = \textstyle\sum_{n=1}^{\infty} \{\cos 2n\pi t\} / \{2n^2\pi^2\} ,$$

it follows readily that

$$\Gamma^2(\alpha+1)e(x) = x^\alpha e^{H(x)} ,$$

$$H(x) = (x+\alpha-\tfrac{1}{2})\log\frac{x+\alpha}{x} - \alpha - \frac{\alpha P(0)}{x(x+\alpha)} + \int_0^\infty P(t)\,\frac{2\alpha(t+x)+\alpha^2}{(t+x)^2(t+x+\alpha)^2}\,dt,$$

and therefore

$$(\tfrac{d}{dx})^k H(x) = O(x^{-1-k}) \qquad (x \to \infty) .$$

Hence

$$\Gamma(\alpha+1)(\tfrac{d}{dx})^k e(x) = x^{\alpha-k}/\Gamma(\alpha+1-k) + O(x^{\alpha-k-1}) .$$

Taking in Lemma 3.17

$$b_0(x) = \Gamma(\alpha+1)e^{([\alpha]+2)}(x),$$

one sees that b_δ coincides with $\Gamma(\alpha+1)u$ and that

$$b_l(x) = \Gamma(\alpha+1)(\tfrac{d}{dx})^{[\alpha]+2-l}e(x) = x^{l-\delta-1}/\Gamma(l-\delta) + O(x^{l-\delta-2})$$

for $l = 0, 1, 2$. Lemma 3.17 then delivers $u(x) = O(x^{-2})$ for $x \to \infty$, and hence $\int|u(t)|dt$ is convergent.

Remark. By Theorem 3.19 it becomes evident that Theorem 3.3 and Theorem 3.18 are optimal in the sense that we have lost nothing of the hypothesis (3.7). Indeed, (3.7) straight-forwardly implies Theorem 3.3, from this Theorem 3.18 follows directly, an application of Theorem 3.18 gives the boundedness of the Riesz means, which implies (3.7) by Theorem 3.19.

By definition, the Riesz means are closely related to the Riemann-Liouville integral (introduced in Lemma 3.17) and therefore especially suited to the calculus of fractional differentiation. In the applications, however, the sequences in question are not as simply structured and the precise evaluation of a suitable fractional derivative may be quite hard.

Therefore it is convenient to derive suffient criteria for a function to belong to $BV_{\alpha+1}$. First observe that most of the functions e given by the examples are smooth at infinity and satisfy $\int_a^\infty x^j|de^{(j)}(x)| < \infty$ for some $j \in \mathbb{N}$ and suitably large $a > 0$. But this condition is sufficient to ensure the existence of fractional derivatives of order $\alpha < j$ for $x > a$, as can easily be seen from the proofs of Lemmata 3.10 - 12, 3.15. In this case one has

$$\text{(3.40)}\quad \begin{aligned} \frac{1}{\Gamma(\alpha+1)}\int_a^\infty x^\alpha|de^{(\alpha)}(x)| &\le \frac{1}{\Gamma(\alpha+1)}\int_a^\infty x^\alpha\frac{1}{\Gamma(j-\alpha)}\int_x^\infty (t-x)^{j-\alpha-1}|de^{(j)}(t)| \\ &\le \frac{1}{\Gamma(j+1)}\int_a^\infty t^j|de^{(j)}(t)| \ . \end{aligned}$$

Our next aim is to develop a sufficient condition for $\int_0^a x^\alpha|de^{(\alpha)}(x)|$ to be finite. This criterium should, however, allow a larger scale of applications than $BV_{j+1} \subset BV_{\alpha+1}$, $\alpha < j$ and $j \in \mathbb{N}$, does. First we suppose e to have a compact support. This is no serious restriction. Indeed, multiply e by a C^∞-function w which is equal to 1 on $[0,a]$ and to zero on $[a+1,\infty)$. As already mentioned, in most examples $\{e(x) - e(x)w(x)\} \in BV_{j+1}$ (though this is not satisfied by the example following Lemma 3.17 taking $j = 1$, $0 < \alpha < 1$), and therefore one has only to show that $ew \in BV_{\alpha+1}$, where ew has compact support. Now, inspired by Weyl's [106] theorem that a function satisfying a Lipschitz condition of order $\beta > 0$ possesses fractional derivatives of order α, $0 < \alpha < \beta$, we are looking for a suitable Lipschitz condition upon $e^{([\alpha+1])}$ which should ensure the existence of $e^{(\alpha+1)}$ and can be verified easily (observe that i) $e \in BV_{\alpha+1}$ always implies the existence of $e^{([\alpha+1])}$ by Lemma 3.15 and ii) that, by applications, we are not interested to cover the case $e \in BV_{\alpha+1}$ with $e^{(\alpha)}$ not locally absolutely continuous).

To this end, suppose $e^{([\alpha+1])}$ to be locally integrable for $x \ge \varepsilon > 0$ and $\int_0^\varepsilon x^\alpha|e^{([\alpha+1])}(x)|dx < \infty$. Thus, since e has compact support, integrals over $e^{([\alpha]+1)}$ will always converge at infinity. Setting $\alpha - [\alpha] = \delta$, $0 < \delta < 1$, suppose further that for $\gamma > \delta$

$$\text{(3.41)}\quad \int_0^\infty x^\alpha|e^{([\alpha+1])}(x+u) - e^{([\alpha+1])}(x)|dx = O(u^\gamma) \qquad (0 < u < 1/4).$$

Then one may integrate $J^{1-\delta}[e^{([\alpha+1])}](x)$ by parts for each $x > 0$ to deduce

$$J^{1-\delta}[e^{([\alpha+1])}](x) = \frac{(y-x)^{-\delta}}{\Gamma(1-\delta)} \int_x^\infty \{e^{([\alpha+1])}(z+y-x)-e^{([\alpha+1])}(z)\}dz\Big|_{y=x}^{\infty}$$

$$- \frac{1}{\Gamma(-\delta)} \int_x^\infty (y-x)^{-\delta-1} \int_x^\infty \{e^{([\alpha+1])}(z+y-x)-e^{([\alpha+1])}(z)\}dzdy \ .$$

Since $0 < \delta < 1$ and since e has compact support, the first term on the right side vanishes at $y = \infty$ for each $x > 0$, and by (3.41) it also vanishes for $y \to x+$, $x > 0$. Furthermore, the second term on the right hand side is absolutely convergent for each $x > 0$ by (3.41), and an interchange of integration yields

$$J^{1-\delta}[e^{([\alpha+1])}](x) = - \frac{1}{\Gamma(-\delta)} \int_x^\infty dz \int_0^\infty u^{-\delta-1}\{e^{([\alpha+1])}(u+z)-e^{([\alpha+1])}(z)\}du.$$

Now the inner integral is a locally integrable function of z for $z > 0$, and therefore $e^{(\alpha+1)}(x)$ exists as a locally integrable function for $x > 0$ by Lemma 3.15 iii). Thus

$$\int_0^\infty x^\alpha|e^{(\alpha+1)}(x)|dx = \int_0^\infty x^\alpha\Big|\frac{1}{\Gamma(-\delta)} \int_0^\infty u^{-\delta-1}\{e^{([\alpha+1])}(x+u)-e^{([\alpha+1])}(x)\}du\Big|dx$$

$$\leq \frac{1}{|\Gamma(-\delta)|} \int_0^\infty u^{-\delta-1} \int_0^\infty x^\alpha|e^{([\alpha+1])}(x+u)-e^{([\alpha+1])}(x)|dx\, du \quad .$$

Since e has compact support and satisfies (3.41) ($\gamma > \delta$), the double integral is convergent. Hence we have proved

<u>*Lemma 3.20.*</u> *Let e have compact support, $e^{([\alpha+1])}$ be locally integrable (excepting the origin) with $\int_0^\varepsilon x^\alpha|e^{([\alpha+1])}(x)|dx$ finite for some $\varepsilon > 0$. If $e^{([\alpha+1])}$ satisfies condition (3.41), then $e^{(\alpha+1)}(x)$ exists almost everywhere and $e \in BV_{\alpha+1}$.*

Let us mention that we have shown that

$$e^{(\alpha+1)}(x)= \frac{-1}{\Gamma(-\delta)} \int_0^\infty u^{-\delta-1}\{e^{([\alpha+1])}(x+u) - e^{([\alpha+1])}(x)\}du \qquad \text{a.e.},$$

the integral being Hadamard's finite part of the fractional Weyl integral (of $e^{([\alpha+1])}$) of negative order $-\delta$; the derivation given here is a modification of Weyl's [106] classical proof.

We want to conclude this section with an analog to Theorem 3.9

("Second theorem of consistency") for the fractional case $\alpha > 0$, $\alpha \notin \mathbb{P}$. To this end we first show

Lemma 3.21. *Let X, $\{P_k\}$ satisfy condition (C^α) (see Def. 3.0) for some $\alpha > 0$, $\alpha \notin \mathbb{P}$. Let $\Phi(t)$ be a non-negative, strictly increasing function with $\lim_{t\to 0+} \Phi(t) = 0$ and $\lim_{t\to\infty} \Phi(t) = \infty$. Furthermore, let there exist $\Phi', \ldots, \Phi^{([\alpha]+2)}$ on $(0,\infty)$ with $\Phi'(t)$ being monotone on $(0,\infty)$ and*

$$(3.42) \qquad t^k |\Phi^{(k+1)}(t)| \leq D\, \Phi'(t) \qquad (0 \leq k \leq [\alpha]+1),$$

D being independent of $t \in (0,\infty)$. Then

$$\| \sum_{k<\rho} (1-\Phi(k)/\Phi(\rho))^\alpha P_k f\| \leq D^*_{\alpha,\Phi} \|f\| \qquad (f \in X).$$

Remark. Lemma 3.21 directly carries over Hardy's "Second theorem of consistency" (for numerical series) to our general framework here. The present proof is based upon the fundamental multiplier Theorem 3.18 and is inspired by proofs of Hirst [64] and Kuttner [68;II] on the same topic. For the sake of simplicity we have assumed stronger hypotheses upon Φ in accordance with Chandrasekharan-Minakshisundaram [39;p.40]; however, these hypotheses are satisfied, e.g., by [6)]

$$(3.43) \qquad t^\alpha (\alpha > 0),\ t^\alpha \log(1+t) (\alpha \geq 0), \{t(t+a)\}^\alpha \quad (\alpha, a > 0).$$

Proof. By Theorem 3.18 we only have to show that [7)]

$$\begin{cases} (1-\Phi(t)/\Phi(\rho))^\alpha, & 0 \leq t \leq \rho \\ 0 & , \quad t \geq \rho \end{cases}$$

6) More precisely, our second example satisfies the monotonicity property of Φ' for all $t > 0$ only for $\alpha \geq 1$; for $0 \leq \alpha < 1$, Φ' is monotonely decreasing for sufficiently large t. But this is sufficient for the above lemma (see e.g. Kuttner [68]) as one may realize by checking the following proof.

7) Note that Kuttner [68;II] has proved (with fewer hypotheses upon Φ) that this condition is necessary and sufficient for Lemma 3.21 to be true in case of numerical series.

belongs to $BV_{\alpha+1}$ uniformly in $\rho > 0$. To this end set

$$(\Phi(\rho)-\Phi(t))^{\alpha} = (\rho-t)^{\alpha}\,[\Phi'(\rho)]^{\alpha} + R(t) \qquad (0 < t < \rho)$$

and $R(t) = 0$ for $t > \rho$. By the proof to Theorem 3.19 (implication (3.36)$\Longrightarrow$(3.37)) it is obvious that

$$\begin{cases} (\rho-t)^{\alpha}, & 0 < t < \rho \\ 0, & t > \rho \end{cases}$$

is bounded in the $BV_{\alpha+1}$-norm by $O(\rho^{\alpha})$. Reasoning as in the proof to Lemma 3.8 it readily follows that $O(\rho^{\alpha}[\Phi'(\rho)]^{\alpha}) = O([\Phi(\rho)]^{\alpha})$. Therefore we have to show that $R(t)/[\Phi(\rho)]^{\alpha} \in BV_{\alpha+1}$ uniformly in ρ. By Lemma 3.15 ii) it is sufficient to prove

$$\|R\|_{BV_{[\alpha]+2}} = O([\Phi(\rho)]^{\alpha}).$$

To this end proceed analogously to the proof of Lemma 3.8 by using the chain rule. First, consider the derivatives of $R(t)$ in the interval $(0,\rho)$. Since $R^{(k)}(\rho) = 0$, $0 \leq k \leq [\alpha]+1$, it is obvious that $R^{([\alpha]+1)}(t)$ is continuous at $t = \rho$. Another differentiation proves $R^{([\alpha]+2)}(t)$ to be a linear combination of the following terms in the interval $(0,\rho)$:

$$I_1(t) = \big(\Phi(\rho)-\Phi(t)\big)^{\alpha-[\alpha]-2}[\Phi'(t)]^{[\alpha]+2} - (\rho-t)^{\alpha-[\alpha]-2}[\Phi'(\rho)]^{\alpha},$$

$$I_k(t) = \big(\Phi(\rho)-\Phi(t)\big)^{\alpha-[\alpha]-3+k}\prod_{n=1}^{k}[\Phi^{(n)}(t)]^{\varepsilon^{n,k}},$$

where $2 \leq k \leq [\alpha]+2$, $\varepsilon^{n,k} \in P$ with (cf. [87;p.20])

$$\sum_{n=1}^{[\alpha]+2}\varepsilon^{n,k} = [\alpha]+3-k, \quad \sum_{n=1}^{[\alpha]+2} n\,\varepsilon^{n,k} = [\alpha]+2 .$$

First estimate $\int t^{[\alpha]+1}|I_k(t)|\,dt$; observing that $\alpha-[\alpha]-3+k > -1$ and proceeding analogously to the proof of Lemma 3.8 one has

$$\int_0^{\rho} t^{[\alpha]+1}|I_k(t)|\,dt = O\Big(\int_0^{\rho}[\Phi(t)]^{[\alpha]+2-k}[\Phi(\rho)-\Phi(t)]^{\alpha-[\alpha]-3+k}\Phi'(t)\,dt\Big)$$

$$= O\Big([\Phi(\rho)]^{[\alpha]+2-k}\int_0^{\rho}[\Phi(\rho)-\Phi(t)]^{\alpha-[\alpha]-3+k}\Phi'(t)\,dt\Big) = O([\Phi(\rho)]^{\alpha}).$$

Hence there remains to estimate $\int_0^\rho t^{[\alpha]+1}|I_1(t)|dt$ suitably. Splitting up $\int_0^\rho = \int_0^{\rho/2} + \int_{\rho/2}^\rho$ it follows that

$$\int_0^{\rho/2} t^{[\alpha]+1}|I_1(t)|dt$$

$$= O\Big(\int_0^{\rho/2}[\Phi(\rho) - \Phi(t)]^{\alpha-[\alpha]-2}[\Phi(t)]^{[\alpha]+1}\Phi'(t)dt\Big)$$

$$+ [\Phi'(\rho)]^\alpha \int_0^{\rho/2} t^{[\alpha]+1}(\rho-t)^{\alpha-[\alpha]-2}dt$$

$$= O\Big([\Phi(\rho/2)]^{[\alpha]+1}\{[\Phi(\rho)-\Phi(\rho/2)]^{\alpha-[\alpha]-1} - [\Phi(\rho)]^{\alpha-[\alpha]-1}\}\Big)$$

$$+ O\Big([\Phi'(\rho)]^\alpha (\rho/2)^{[\alpha]+1}\{(\rho/2)^{\alpha-[\alpha]-2} - \rho^{\alpha-[\alpha]-1}\}\Big)$$

$$= O\Big([\Phi(\rho)]^\alpha\Big) + O\Big([\Phi(\rho/2)]^{[\alpha]+1}[\tfrac{\rho}{2}\Phi'(\rho_1)]^{\alpha-[\alpha]-1}\Big) = O([\Phi(\rho)]^\alpha)$$

($\rho/2 < \rho_1 < \rho$ being given by an application of the mean value theorem) on account of the monotonicity properties of Φ and Φ'.

Concerning the last integral $\int_{\rho/2}^\rho$ we first apply the mean value theorem (of differentiation) twice to

$$I_\rho(t) \equiv [\frac{\Phi(\rho)-\Phi(t)}{\rho - t}]^{\alpha-[\alpha]-2}[\Phi'(t)]^{[\alpha]+2} - [\Phi'(\rho)]^\alpha$$

$$= O((\rho-t)\{[\Phi'(t_1)]^{\alpha-[\alpha]-3}|\Phi''(t_1)|[\Phi'(t)]^{[\alpha]+2}+[\Phi'(t_2)]^{\alpha-1}|\Phi''(t_2)|\})$$

where $t < t_1 < \rho$ and $t < t_2 < \rho$, to obtain

$$\int_{\rho/2}^\rho t^{[\alpha]+1}|I_1(t)|dt = \int_{\rho/2}^\rho t^{[\alpha]+1}(\rho-t)^{\alpha-[\alpha]-2}|I_\rho(t)|dt$$

$$= O\Big(\int_{\rho/2}^\rho t^{[\alpha]}(\rho-t)^{\alpha-[\alpha]-1}\max\{[\Phi'(\rho)]^\alpha,[\Phi'(\rho/2)]^\alpha\}\,dt\Big)$$

$$= O\Big((\rho/2)^\alpha \max\{[\Phi'(\rho)]^\alpha, [\Phi'(\rho/2)]^\alpha\}\Big) = O([\Phi(\rho)]^\alpha)$$

again by the monotonicity properties of Φ and Φ'. Thus the lemma is proved completely.

Now the analog to Theorem 3.9 for fractional α is an easy consequence of the above lemma.

Theorem 3.22. *Let* X, $\{P_k\}$ *satisfy condition* (C^α) *for some* $\alpha > 0$, $\alpha \notin P$; *let* Φ *be as in Lemma 3.21. If, furthermore,* $\Psi(\rho) > 0$ *for* $\rho > 0$ *and* $e(t) \in BV_{\alpha+1}$, *then*

$$\eta(\rho) = \{ e(\Phi(k)/\Psi(\rho))\}_{k=0}^{\infty} \in M(X;\{P_k\})$$

uniformly in $\rho > 0$.

Proof. Since by hypothesis

$$\int_0^\infty (\Phi(u)/\Psi(\rho))^\alpha |de^{(\alpha)}(\Phi(u)/\Psi(\rho))| = O(1) \qquad (\rho > 0),$$

we conclude by the above lemma that the Bochner integral

$$(-1)^{[\alpha]+1} f^\eta = \int_0^\infty \sum_{k<u} \left(1-\frac{\Phi(k)}{\Phi(u)}\right)^\alpha P_k f \; (\Phi(u)/\Psi(\rho))^\alpha de^{(\alpha)}(\Phi(u)/\Psi(\rho))$$

exists in the X-norm, and thus defines an element $f^\eta \in X$ with

$$\|f^\eta\| \leq C\|e\|_{BV_{\alpha+1}} \|f\| .$$

Substituting $\Phi(u)/\Psi(\rho) = t$ and applying P_n upon f^η yield

$$\begin{aligned}(-1)^{[\alpha]+1} P_n f^\eta &= P_n \int_0^\infty \sum_{\Phi(k)<t\Psi(\rho)} \left(t - \frac{\Phi(k)}{\Psi(\rho)}\right)^\alpha P_k f \; de^{(\alpha)}(t) \\ &= P_n f \int_{\Phi(n)/\Psi(\rho)}^\infty \left(t - \frac{\Phi(n)}{\Psi(\rho)}\right)^\alpha de^{(\alpha)}(t) \\ &= (-1)^{[\alpha]+1} e(\Phi(n)/\Psi(\rho)) P_n f\end{aligned}$$

by Lemma 3.14, and so Theorem 3.22 is proved completely.

Having established a multiplier theory for (C,α)-bounded expansions we proceed to test this theory for particular summation processes in the next section.

4. PARTICULAR SUMMATION METHODS

As already mentioned it is assumed throughout that the (C,α)-means of the expansion $\sum P_k f$ are uniformly bounded for some $\alpha \geq 0$. In 4.1, an approximation process is introduced which does not satisfy condition (F) (cf. Def. 2.2); further, a general Bernstein-type inequality is derived on the direct sum of $\{P_k(X)\}_{k=0}^n$, i.e., on

$$\bigoplus_{k=0}^{n} P_k(X) = \{f \in X;\ f = \textstyle\sum_{k=0}^{n} P_k f\ \};$$

in particular, this yields Bernstein-type inequalities for all polynomial operators. Finally, we discuss the inverse operator $[B^{\Phi}]^{-1}$ of B^{Φ} which occures in the Bernstein-type inequality.

In Sec. 4.2, the general theory of Sec. 2 and the multiplier criteria of Sec. 3 are applied to the Abel-Cartwright means. The evaluations are quite simple in contrast to Sec. 4.3 which is devoted to Riesz means of fractional order; in particular the results of Sec. 3.3 are basic . Finally in the last subsections Jackson- and Bernstein-type inequalities are derived for the Bessel potentials as well as for the Cesàro and de La Vallée-Poussin means. The results of [33;I,II], [59;I] in the integer case are contained in the present treatment and are now developed further; in the fractional case the results are naturally new.

4.1 A polynomial summation method; a Bernstein-type inequality for polynomials

Consider a C^{∞}-function $t(x)$ on $[0,\infty)$, which equals 1 for $0 \leq x \leq 1$ and equals 0 for $x \geq 2$. Clearly $t \in BV_{j+1}$ for each $j \in P$, and thus $\{\tau(\rho)\} = \{\{t(k/\rho)\}_{k=0}^{\infty}\}$ is a summation method which does not satisfy condition (F). Since obviously $(t(x)-1)/x \in BV_{j+1}$, $j \in P$, one has in particular (choose $\Phi(x) = x^{\beta}$, $\Psi(\rho) = \rho^{\beta}$, $\beta > 0$ in Lemma 3.8) by Theorems 2.1, 3.9, and Lemma 3.15 ii)

Theorem 4.1. *Let X, $\{P_k\}$ satisfy condition (C^{α}) (see Def. 3.0) for some $\alpha \geq 0$. The approximation process $\mathcal{T}$, defined by the associated multiplier sequences $\tau_k(\rho) = t\big((k/\rho)^{\beta}\big)$ with t as specified above, does not satisfy condition (F). Furthermore, for arbitrary $\beta > 0$ one has the Jackson-type inequality*

$$\|T(\rho)f-f\| \leq D\ \rho^{-\beta}|f|_{\psi} \qquad (f \in X^{\psi})$$

with $\psi = \{k^{\beta}\}$.

Theorem 2.1 b) calls for a Bernstein-type inequality. We settle this problem by deriving a Bernstein-type inequality for arbitrary $f \in \bigoplus_{k=0}^{n} P_k(X)$. To this end we observe that by Taylor's two point interpolation theorem (cf.[42;p.37]) there exists a unique polynomial of degree (2j+3) satisfying

$$p_{2j+3}^{(k)}(1) = \begin{cases} 1, & k = 0,1 \\ 0, & 2 \leq k \leq j+1 \end{cases}, \quad p_{2j+3}^{(k)}(2) = \begin{cases} 2, & k = 0 \\ 0, & 1 \leq k \leq j+1 \end{cases}.$$

Now consider the function

$$e(x) = \begin{cases} x & , 0 \leq x \leq 1 \\ p_{2j+3}(x), & 1 \leq x \leq 2 \\ 2 & , x \geq 2 \end{cases};$$

by the choice of p_{2j+3}, all derivatives $e^{(k)}(x)$, $0 \leq k \leq j+1$, are continuous functions. Hence $(j \geq 2)$

$$\int_0^{\infty} x^j |e^{(j+1)}(x)|\,dx = \int_1^2 x^j |p_{2j+3}^{(j+1)}(x)|\,dx < \infty ,$$

and the family $\{\eta(n)\}_{n\in \mathbb{N}} \subset s$, $\eta_k(n) = e(k/n)$ for $k \in \mathbb{P}$, is a uniformly bounded multiplier family by Theorem 3.7 and Lemma 3.15 ii) ($\lim_{x\to\infty} e(x) = 2$ causes no difficulties, for one may consider $e(x) - 2$ and this argumentation is always supposed tacitly). Finally, an application of Theorem 3.9 yields

Theorem 4.2. *Let* X, $\{P_k\}$ *satisfy condition* (C^{α}) *for some* $\alpha \geq 0$. *Let* $\Phi(x)$ *be a non-negative, strictly increasing function with* $\lim_{x\to 0+} \Phi(x) = 0$ *and* $\lim_{x\to\infty} \Phi(x) = \infty$; *let* Φ *possess continuous derivatives of order* $(j+1)$ $(j \in \mathbb{P},\ j > \alpha)$ *on* $(0,\infty)$ *with*

$$(4.1) \qquad |x^r\ \Phi^{(r+1)}(x)| \leq D\ \Phi'(x) \qquad (0 < x < \infty,\ 0 \leq r \leq j) .$$

Then the restriction of B^{Φ} *to* $\bigoplus_{k=0}^{n} P_k(X)$ *satisfies*

$$\|\textstyle\sum_{k=0}^{n} \Phi(k) P_k f\| \leq D^{*}\ \Phi(n) \|\sum_{k=0}^{n} P_k f\| \qquad (f \in X),$$

the constant D^{*} *being independent of* n *and* f. *In particular,*

$$\{(k(k+\omega))^{\beta}\}_{k\in P} \quad (\beta > 0, \omega \geq 0), \quad \{k^{\alpha}\log(1+k)\}_{k\in P} \quad (\alpha \geq 0)$$

are admitted.

Remark. In the case $\alpha = 1$ this theorem is already proved in [59;I]. The above method of proof may also be applied to derive an example of a non-polynomial summation method satisfying the assertions of Theorem 4.1.

By the way, Theorem 4.1 itself supplies us with some type of inverse; for, taking $\rho = n/2$ one obtains $(\Phi(k) = \psi_k)$

$$\| f\| \leq D_1^* [\Phi(n)]^{-1} |f|_{\psi}$$

for all $f \in X / \bigoplus_{k=0}^{n} P_k(X)$, i.e., $P_k f = 0$ for $0 \leq k \leq n$.

We conclude this subsection with a discussion of the inverse of B^{Φ} which, in concrete examples, turns out to be some fractional integral operator.

Since $\Phi(0) = 0$ it is obvious that the domain of $[B^{\Phi}]^{-1}$ necessarily has to be a subset of $X / P_0(X)$.

Lemma 4.3. Let X, $\{P_k\}$, Φ be as in Theorem 4.2. Then $[B^{\Phi}]^{-1}$ with associated multiplier sequence $\{1/\Phi(k)\}$ is continuous on $X / P_0(X)$.

For the proof note that by Theorem 3.9, Lemma 3.15 ii) and (3.20) one has only to show that $\int_1^{\infty} x^j |(d/dx)^{j+1} x^{-1}| dx < \infty$ for some $j \geq \alpha$ which is fairly obvious.

Remark. What is bothering about Lemma 4.3 is the fact that not the whole space X is admitted as domain of $[B^{\Phi}]^{-1}$. But this may be avoided in a certain sense by a trick which arises from a lemma of Stein [95] (cf. also [29],[102]):

Suppose there are multiplier sequences $\mu^{(1)}, \mu^{(2)}, \mu^{(3)} \in M$ and a closed linear operator B^{ψ} with continuous inverse on all of X such that

$$\begin{aligned} \Phi(k) &= \mu_k^{(1)} \psi_k \\ \psi_k &= \mu_k^{(2)} + \Phi(k)\mu_k^{(3)} \end{aligned} \qquad (k \in P) . \tag{4.2}$$

Then the domains of B^{Φ} and B^{ψ} are identical with equivalent norms. Indeed, let $f \in X^{\psi}$; then

$$\|f\|_{\Phi} = \|f\| + \|B^{\Phi}f\| \leq \|f\| + \|\mu^{(1)}\|_M \|B^{\psi}f\| \leq D_1 \|f\|_{\psi}$$

and analogously the converse direction. Thus the domain of B^{Φ} coincides with $[B^{\psi}]^{-1}(X)$. It remains to show that the latter space is independent of a particular ψ whenever ψ satisfies (4.2). Let φ, $\nu^{(i)} \in M$ also satisfy (4.2); then it readily follows from (4.2) that $\{\psi_k/\varphi_k\}$, $\{\varphi_k/\psi_k\} \in M$ and therefore $[B^{\psi}]^{-1}(X) = [B^{\varphi}]^{-1}(X)$ under suitable norms. Finally note that $\{\psi_k\} = \{1 + \Phi(k)\}_{k \in P}$ is an admissible (but not unique) choice.

4.2 Abel-Cartwright means

Let us introduce the approximation process $\mathfrak{W}_{\Phi}$ for $\rho > 0$ via

$$(4.3) \qquad W_{\Phi}(\rho)f \sim \sum_{k=0}^{\infty} w(\Phi(k)/\Phi(\rho))P_k f, \qquad w(x)=e^{-x},\ x \geq 0,$$

where Φ is supposed to satisfy the conditions of Lemma 3.8.

In particular, $\Phi(k) = k^{\kappa}, \kappa > 0$, yields the standard Abel-Cartwright means, with equality holding in (4.3) on account of (3.14). To show that $\mathfrak{W}_{\Phi}$ really is a summation method, in view of Theorem 3.9 and Lemma 3.15 ii) one only has to show that $e^{-x} \in BV_{j+1}$, $j \in P$, $j \geq \alpha$, and $\lim_{\rho\to\infty} w(\Phi(k)/\Phi(\rho)) = 1$ (Banach-Steinhaus theorem) which are both quite obvious. Thus we may apply the results of Sec. 2 and 3 to derive

Theorem 4.4. *Let X, $\{P_k\}$ satisfy condition (C^{α}) for some $\alpha \geq 0$, let Φ be as in Theorem 4.2, and $\mathfrak{W}_{\Phi}$ be given by (4.3). Then with $\psi_k = (\Phi(k))^{\gamma}$*

i) $$\|W_{\Phi}(\rho)f-f\| \leq D_1[\Phi(\rho)]^{-\gamma}|f|_{\psi} \qquad (f \in X^{\psi})$$

for $0 < \gamma \leq 1$;

ii) $$\|B^{\psi} W_{\Phi}(\rho)f\| \leq D_2[\Phi(\rho)]^{\gamma}\|f\| \qquad (f \in X)$$

***for all** $\gamma > 0$;*

***iii)** a)* $$\|\Phi(\rho)\{W_{\Phi}(\rho)f-f\}\| = o(1) \qquad (\rho \to \infty)$$

implies $f \in P_0(X)$ and $W_{\Phi}(\rho)f = f$ for all $\rho > 0$;

b) the Favard class of $\mathfrak{W}_{\Phi}$ is the set $(X^{\psi})^{\sim X}$ where $\psi_k = \Phi(k)$, thus

$\gamma = 1$, and

$$\sup_{\rho>0} \Phi(\rho)\| W_\Phi(\rho)f-f\| \sim |f|_{\psi^\sim}$$

$$\sim \sup_{n\in\mathbb{N}} \| \sum_{k=0}^{n} (A_{n-k}^{\alpha} / A_n^{\alpha})\Phi(k)P_k f\|$$

are equivalent semi-norms on $(X^\psi)^{\sim X}$;

iv) $$\| B^\psi W_\Phi(\rho)f\| \leq D_3[\Phi(\rho)]^\gamma \| W_\Phi(\rho)f-f\| \qquad (f \in X)$$

for $\gamma \geq 1$

v) $$\| W_{[\Phi]^\gamma}(\rho)f-f\| \leq D_4 \| W_\Phi(\rho)f-f\| \qquad (f \in X)$$

for all $\gamma > 1$.

Proof. On account of Theorems 2.1, 2.6, 2.7, 2.8, 3.9 and Lemmata 2.3 and 3.15 ii) we only have to verify that the functions

$$e_1(x) = x^{-\gamma}(e^{-x}-1) \quad (0 < \gamma \leq 1), \quad e_2(x) = x^\gamma e^{-x} \quad (\gamma > 0),$$

$$e_3(x) = \frac{x^\gamma e^{-x}}{1-e^{-x}} \quad (\gamma \geq 1), \qquad e_4(x) = \frac{1-e^{-x^\gamma}}{1-e^{-x}} \quad (\gamma > 1)$$ [8)]

belong to some BV_{j+1} with $j \geq \alpha$ (condition (F) (see Def. 2.2) is trivially satisfied since $\lim_{x\to 0} x^{-1}(1-e^{-x}) = 1$ and $K = \{0\}, \Phi$ being strictly increasing). Now $e_1^{(j+1)}(x)$ obviously exists for all $x > 0$, $j \in P$ and $\int_1^\infty x^j |e^{(j+1)}(x)|dx < \infty$. Hence only the behaviour of $e^{(j+1)}(x)$ at the origin is of interest. Observing that

$$e_1'(x) = \begin{cases} O((1-\gamma)x^{-\gamma}), & 0 < \gamma < 1 \\ O(1), & \gamma = 1, \end{cases}$$

and setting $e_1'(x) = Z(x)/x^{\gamma+1}$ it follows that for arbitrary $k \in P$

$$Z^{(k)}(x) = O(x^{1-k}) \quad (0 < \gamma < 1), \quad Z^{(k)}(x) = O(x^{2-k}) \quad (\gamma = 1).$$

---

8) Since $\lim_{x\to\infty} e_4(x) = 1$, we should have to consider $e_4(x) - 1$; but obviously the constant causes no difficulties and we may omit it - now and in the following.

Hence Leibniz' rule gives for $x \to 0+$

$$e_1^{(j+1)}(x) = \sum_{m=0}^{j} \binom{j}{m} Z^{(j-m)}(x)\left(\frac{d}{dx}\right)^m x^{-\gamma-1} = \begin{cases} O(x^{-\gamma-j}), 0<\gamma<1 \\ O(x^{-j}), \quad \gamma = 1, \end{cases}$$

and thus $\int_0^1 x^j |e_1^{(j+1)}(x)|dx < \infty$, i.e., $e_1 \in BV_{j+1}$ for each $j \in P$.

To show that $e_2 \in BV_{j+1}$ is still simpler. Concerning e_3 note that $\gamma \geq 1$ is important to neutralize the singularity at $x = 0$ caused by the denominator; the computations follow along the above lines and give $e_3 \in BV_{j+1}$, $j \in P$.

Though the computations for $e_4 \in BV_{j+1}$ are not difficult, we sketch them here. Differentiation of e_4 yields $(x > 0)$

$$e_4'(x) = \frac{\gamma\, x^{\gamma-1} e^{-x^\gamma}}{(1-e^{-x})} - \frac{1-e^{-x^\gamma}}{(1-e^{-x})^2}\, e^{-x} = \frac{Z(x)}{(1-e^{-x})^2} \quad ,$$

and repeated differentiation gives $e^{(j+1)}(x) = O(e^{-x} + (\gamma\, x^{\gamma-1})^{j+1} e^{-x^\gamma})$ for $x \to \infty$; thus $\int_1^\infty x^j |e_4^{(j+1)}(x)|dx < \infty$. Hence only the behaviour of e_4 at the origin is of interest. An examination of $Z(x)$ gives $Z^{(k)}(x) = O(x^{\delta-k})$, $1 < \delta = \min\{\gamma, 2\}$, for $x \to 0+$ and all $k \in P$. Thus Leibniz' rule yields

$$e_4^{(j+1)}(x) = \sum_{k=0}^{j} \binom{j}{m} Z^{(j-m)}(x)\left(\frac{d}{dx}\right)^m (1-e^{-x})^{-2} = O(x^{\delta-2-j})$$

for $x \to 0+$ since $(d/dx)^m (1-e^{-x})^{-2} = O(x^{-2-j})$, $0 \leq m \leq j$.

Since $\delta > 1$ one finally arrives at

$$\int_0^1 x^j |e^{(j+1)}(x)|dx = O\left(\int_0^1 x^{\delta-2}\, dx\right) = O(1).$$

Thus the theorem is completely proved.

For $\Phi(k) = k^\kappa$, $\kappa > 0$, statement ii) is already proved in [59;I] iii) in [33;II], v) in [33;I] for $\alpha = 1$; furthermore iii) with $\Phi(k) = k(k+c)$, $c > 0$, is to be found in [33;II].

Since the choice $\Phi(k) = (k(k+c))^{\kappa/2}$ will be important for expansions into Jacobi polynomials and spherical harmonics, we show for the sake of completeness

Lemma 4.5. *Let X, $\{P_k\}$ satisfy condition (C^{α}) for some $\alpha \geq 0$. Under the choice $\psi = \{k^{\kappa}\}$, $\psi^* = \{(k(k+c))^{\kappa/2}\}$, $c, \kappa > 0$, the spaces $(X^{\psi})^{\sim X}$ and $(X^{\psi^*})^{\sim X}$ coincide with equivalent semi-norms.*

Proof. It is sufficient to prove that $\eta(\beta) \in bv_{j+1}$ for each $j \in P$ and $\beta \in R$, where

$$\eta_k(\beta) = \begin{cases} 0 & , k = 0 \\ (k/(k+c))^{\beta}, & k \in N \end{cases} .$$

If $\beta > 0$ ($\beta = 0$ is trivial) the function $e(x) = (x/(x+c))^{\beta}$ obviously belongs to BV_{j+1} (cf. Footnote 8), and hence by Theorem 3.7 $\eta(\beta) \in bv_{j+1}$. In case $\beta < 0$ consider

$$\|\eta(\beta)\|_{bv_{j+1}} = 1 + |\Delta^{j+1}\, \eta_0(\beta)| + \sum_{k=1}^{\infty} A_k^j\, |\Delta^{j+1}\eta_k(\beta)|$$

and observe that the latter sum may be estimated by $1/j!\, \int_1^{\infty} x^j |e^{(j+1)}(x)|\,dx < \infty$ on account of (3.20). Since furthermore $\Delta^{j+1}\, \eta_0(\beta)$ consists of only finitely many terms and thus is bounded, the lemma is completely established.

The above lemma is already contained in [33;II] for $\kappa = 2$.

4.3 Riesz means

Let us introduce the Riesz means $\mathcal{R}_{\Phi,\lambda}$ of order $\lambda > 0$ for $\rho > 0$ by

$$(4.4) \qquad R_{\Phi,\lambda}(\rho)f = \sum_{k<\rho} r(\Phi(k)/\Phi(\rho))P_k f, \quad r(x) = \begin{cases} (1-x)^{\lambda}, & 0 \leq x \leq 1 \\ 0, & x \geq 1 \end{cases} ,$$

where Φ is supposed to satisfy the conditions of Lemma 3.21. Clearly, $\lim_{\rho\to\infty} r(\Phi(k)/\Phi(\rho)) = 1$ and therefore $\mathcal{R}_{\Phi,\lambda}$ is a summation method for $\lambda \geq \alpha$ by Lemma 3.21 and the Banach-Steinhaus theorem. First we establish

Theorem 4.6. *Let X, $\{P_k\}$ satisfy condition (C^{α}) for some $\alpha \geq 0$; let Φ be as in Theorem 4.2 - Φ satisfying (4.1) for $0 \leq k \leq \alpha$ provided α is*

an integer and for $0 \leq k \leq [\alpha] + 1$ provided $\alpha \notin P$; in the fractional case $\alpha \notin P$ assume further Φ' to be monotone on $(0,\infty)$. If furthermore $0 \leq \alpha < \lambda < \infty$ provided $\alpha \notin N$, and $\alpha \leq \lambda$ provided $\alpha \in N$, then the Abel-Cartwright means $\mathfrak{W}_\Phi$ of (4.3) and the Riesz means $\mathcal{R}_{\Phi,\lambda}$ of (4.4) are equivalent in the sense of Def. 1.4, i.e.,

$$\| W_\Phi(\rho)f-f\| \approx \| R_{\Phi,\lambda}(\rho)f-f\| \qquad (f \in X).$$

Proof. By Theorems 3.9 and 3.22 we have to verify that the functions $e(x)-1$, $[e(x)]^{-1} - 1$ are bounded in the $BV_{\alpha+1}$-norm, where

$$e(x) = \begin{cases} \dfrac{1-e^{-x}}{1-(1-x)^\lambda}, & 0 \leq x \leq 1 \\ 1 - e^{-x}, & x \geq 1 \end{cases}, \quad [e(x)]^{-1} = \begin{cases} \dfrac{1-(1-x)^\lambda}{1-e^{-x}}, & 0 < x < 1 \\ (1-e^{-x})^{-1}, & x \geq 1 \end{cases}.$$

It is easy to see that the behaviour of $e(x)$, $[e(x)]^{-1}$, respectively, is of interest only at the origin, at $x = 1$, and at infinity. Moreover, since $\int_{1+}^\infty x^{[\alpha]+1}|e^{([\alpha]+2)}(x)|dx < \infty$ (and analogously for $[e(x)]^{-1}$), by (3.40) one has to show that $\int_0^{1+} x^\alpha|de^{(\alpha)}(x)| < \infty$ (and analogously for $[e(x)]^{-1}$).

First consider $e(x)$ and distinguish the cases $\alpha \in P$ and $\alpha \notin P$.

A. Let $\alpha \in P$, $\lambda \geq \alpha$, $\lambda > 0$ and consider the behaviour of $e(x)$ at the origin. Setting $e'(x) = Z(x)/(1-(1-x)^\lambda)^2$ and observing that $\lim_{x\to 0+} e'(x) = (\lambda^2 - 2\lambda)/2\lambda^2$ it follows that

$$Z^{(k)}(x) = O((1 - (1-x)^\lambda)^{2-k}) \qquad (k \in P,\ x \to 0+).$$

Leibniz' rule yields again

$$(4.5) \qquad e^{(k+1)}(x) = \sum_{m=0}^{k} \binom{k}{m} Z^{(k-m)}(x)\left(\frac{d}{dx}\right)^m (1-(1-x)^\lambda)^{-2}$$

$$= O((1-(1-x)^\lambda)^{-k}) = O(x^{-k}) \qquad (k \in P)$$

and hence $\int_0^{1/2} x^\alpha|e^{(\alpha+1)}(x)|dx < \infty$ if $\alpha \in P$.

Now let us consider the behaviour of $e(x)$ at $x = 1$ in case $\alpha = \lambda \in N$. Leibniz' rule again gives for $0 \leq k \leq \lambda$

(4.6) $$e^{(k)}(x) = \sum_{m=o}^{k} \binom{k}{m} (\tfrac{d}{dx})^{k-m}(1-e^{-x})(\tfrac{d}{dx})^{m}(1-(1-x)^{\lambda})^{-1} .$$

Since $(1-(1-x)^{\lambda})^{-k}$ and derivatives of $(1-e^{-x})$ remain uniformly bounded in $(1/2, 1)$, the only possible singular behaviour of $e^{(k)}(x)$ at $x = 1$ results from a repeated differentiation of $\lambda(1-x)^{\lambda-1}$ (this term arising from an application of the chain rule to $(1-(1-x)^{\lambda})^{-1}$). In the most disadvantageous case there arises an expression of type $O((1-x)^{\lambda-k})$, and hence $e^{(\lambda)}(1-) = O(1)$. Since $e^{(\lambda)}(1+) = O(1)$, $e^{(\lambda)}(x)$ has at most a finite jump at $x = 1$ ($e^{(k)}(x)$ is continuous at $x = 1$ for $0 \leq k \leq \lambda - 1$). Thus $\int_{1-}^{1+} x^{\alpha}|de^{(\alpha)}(x)|$ is bounded. Furthermore, $e^{(\alpha+1)}(x) = e^{(\lambda+1)}(x)$ exists as a bounded function in $(1/2,1)$, and therefore $\int_{1/2}^{1-} x^{\alpha}|e^{(\alpha+1)}(x)|dx < \infty$.

Concerning the case $\lambda > \alpha \in \mathbf{P}$ observe that $e^{(\alpha)}(x)$ is continuous at $x = 1$ (which is readily seen from (4.6)) and that $e^{(\alpha+1)}(x) = O((1-x)^{\lambda-\alpha-1})$ for $x \to 1-$. Thus $\int_{1/2}^{1} x^{\alpha}|e^{(\alpha+1)}(x)|dx < \infty$, and case A is established.

B. Now consider $0 < \alpha < \lambda$, $\alpha \notin \mathbf{P}$. By Lemma 3.15 ii) we may suppose without loss of generality that $[\alpha] = [\lambda] < \alpha < \lambda$. We wish to apply Lemma 3.20. In part A of this proof it is already shown (implicitly) that $\int_{o}^{\varepsilon} x^{\alpha}|e^{([\alpha+1])}(x)|dx$ is finite when $\varepsilon \leq 1/2$, so that (by the argument following (3.40)) we have only to verify

(4.7). $$I(u) \equiv \int_{o}^{3/2} x^{\alpha}|e^{([\alpha]+1)}(x+u)-e^{([\alpha]+1)}(x)|dx = O(u^{\lambda-[\lambda]})$$

for $0 < u < 1/4$. To this end, split $I(u)$ into five parts:

(4.8) $$I(u) \equiv \int_{o}^{3/2} \ldots = (\int_{o}^{1/2} + \int_{1/2}^{1-2u} + \int_{1-2u}^{1-u} + \int_{1-u}^{1} + \int_{1}^{3/2})\ldots \equiv I_1+I_2+I_3+I_4+I_5.$$

To estimate I_1 use (4.5) for $k = [\alpha] + 1$ to deduce

$$I_1 \leq \int_{o}^{1/2} x^{\alpha} \int_{x}^{x+u} |e^{([\alpha]+2)}(v)|dvdx = O(u\int_{o}^{1/2} x^{\alpha-[\alpha]-1}dx) = O(u) .$$

To estimate I_5 note that $e^{([\alpha]+2)}(x) = \pm e^{-x}$ for $x > 1$, and hence

$$I_5 \leq \int_{1}^{3/2} x^{\alpha} \int_{x}^{x+u} e^{-v}dvdx \leq u \int_{1}^{3/2} x^{\alpha}e^{-x}dx = O(u).$$

Analogously one can estimate I_2 by observing that $e^{([\alpha]+2)}(v) = O((1-v)^{\lambda-[\alpha]-2})$, $1/2 \leq v < 1$; indeed,

$$I_2 \leq \int_{1/2}^{1-2u} x^{\alpha} \int_{x}^{x+u} |e^{([\alpha]+2)}(v)|dvdx$$

$$= O\Big(\int_{1/2}^{1-2u} x^{\alpha}\{(1-x-u)^{\lambda-[\alpha]-1} - (1-x)^{\lambda-[\alpha]-1}\}dx\Big)$$

$$= O(u^{\lambda-[\alpha]}) + O\Big((\tfrac{1}{2}-u)^{\lambda-[\alpha]} - (\tfrac{1}{2})^{\lambda-[\alpha]}\Big) = O(u^{\lambda-[\alpha]}).$$

$$I_3 \leq \int_{1-2u}^{1-u} x^{\alpha}\{|e^{([\alpha]+1)}(x+u)| + |e^{([\alpha]+1)}(x)|\}dx$$

$$\leq \int_{1-2u}^{1} x^{\alpha}|e^{([\alpha]+1)}(x)|dx = O\Big(\int_{1-2u}^{1} (1-x)^{\lambda-[\alpha]-1}dx\Big) = O(u^{\lambda-[\alpha]}).$$

Finally,

$$I_4 \leq \int_{1-u}^{1} |e^{([\alpha]+1)}(x)|dx + \int_{1}^{1+u} |e^{([\alpha]+1)}(x)|dx$$

$$= O\Big(\int_{1-u}^{1} (1-x)^{\lambda-[\alpha]-1}dx\Big) + \int_{1}^{1+u} e^{-x}\,dx = O(u^{\lambda-[\alpha]}).$$

Combining the single estimates for I_k one obtains (4.7) and hence $e^{(\alpha+1)}(x)$ exists almost everywhere and $\{e(x) - 1\} \in BV_{\alpha+1}$ in view of Lemma 3.20 and (3.40).

Analogously one shows $\{[e(x)]^{-1} - 1\} \in BV_{\alpha+1}$, where the behaviour of $[e(x)]^{-1}$ at the origin and at $x = 1$ is of a similar structure to that of $e(x)$. Hence the theorem is established.

In case $\Phi(k) = k^{\kappa}$, $\kappa > 0$, $\alpha = \lambda = 1$, this theorem is proved in [33;I]. Naturally it is annoying that in the above theorem the case $\alpha = \lambda$ for $\alpha \notin \mathbb{N}$ is excluded; very precise estimates of $e(x)$ (and $[e(x)]^{-1}$) should also deliver this case (where $e^{(\alpha)}(x)$ will have a finite jump at $x = 1$). Nevertheless, in comparison with the result in [33;I] the present extension is rather farreaching and easily allows one to prove the following analog of Theorem 4.4.

<u>*Theorem 4.7.*</u> *Let X, $\{P_k\}$, Φ, α, λ be as in Theorem 4.6, and let*

$\mathcal{R}_{\Phi,\lambda}$ be given by (4.4). Then with $\psi_k = (\Phi(k))^\gamma$

i) $$\|R_{\Phi,\lambda}(\rho)f-f\| \leq D_1(\Phi(\rho))^{-\gamma}|f|_\psi \qquad (f \in X^\psi)$$

for $0 < \gamma \leq 1$;

ii) $$\|B^\psi R_{\Phi,\lambda}(\rho)f\| \leq D_2(\Phi(\rho))^\gamma \|f\| \qquad (f \in X)$$

for all $\gamma > 0$;

iii) a) $$\Phi(\rho)\|R_{\Phi,\lambda}(\rho)f-f\| = o(1) \qquad (\rho \to \infty)$$

implies $f \in P_o(X)$ and $R_{\Phi,\lambda}(\rho)f = f$ for all $\rho > 0$;

b) the Favard class of $\mathcal{R}_{\Phi,\lambda}$ is the set $(X^\psi)^{\sim X}$ where $\psi_k = \Phi(k)$, thus $\gamma = 1$, and

$$\sup_{\rho>0} \Phi(\rho)\|R_{\Phi,\lambda}(\rho)f-f\| \sim |f|_{\psi^\sim}$$

$$\sim \sup_{n\in\mathbb{N}}\|\sum_{k=o}^{n}(A_{n-k}^{\alpha}/A_n^{\alpha})\ \Phi(k)P_k f\|$$

are equivalent semi-norms on $(X^\psi)^{\sim X}$;

iv) $$\|B^\psi R_{\Phi,\lambda}(\rho)f\| \leq D_3(\Phi(\rho))^\gamma \|R_{\Phi,\lambda}(\rho)f-f\| \qquad (f \in X)$$

for $\gamma \geq 1$;

v) $$\|R_{[\Phi]^\gamma,\lambda}(\rho)f-f\| \leq D_4\|R_{\Phi,\lambda}(\rho)f-f\| \qquad (f \in X)$$

for all $\gamma > 1$.

Proof. i), iii), v) immediately follow by Theorems 4.4, 4.6. To proof ii) and iv) one has only to establish

$$\|B^\psi R_{\Phi,\lambda}(\rho)f\| \leq D_5\|B^\psi W_\Phi(\rho)f\| \qquad (f \in X) . \tag{4.9}$$

By the technique of Sec. 2 and Theorems 3.9 and 3.22 one only has to verify that

$$e(x) = \begin{cases} (1-x)^\lambda e^x, & 0 \leq x \leq 1 \\ 0 \quad , & x \geq 1 \end{cases}$$

belongs to $BV_{\alpha+1}$, with α,λ as specified in Theorem 4.6. Obviously, e(x) and all its derivatives are bounded at the origin. Thus the behaviour of $e^{(k)}(x)$ at $x = 1$ has to be discussed. In case $\lambda \geq \alpha \in P$, $\lambda > 0$, $e \in BV_{\alpha+1}$ follows very simply since the derivative $e^{(\alpha)}(x)$ has a finite jump at $x = 1$ provided $\lambda = \alpha$, or $e^{(\alpha+1)}(x)$ is locally integrable at $x = 1$ provided $\lambda > \alpha$.

In case $\lambda > \alpha \notin P$ we again apply Lemma 3.20 where, without loss of generality, we may suppose $[\lambda] = [\alpha] < \alpha < \lambda$. We have to verify (3.41) and split the arising expression into five terms analogous to (4.8). Obviously $I_1 = O(u)$ and $I_5 = O(u)$ for $0 < u < 1/4$. Furthermore, by Leibniz' rule

$$e^{(k)}(x) = \sum_{m=0}^{k} \binom{k}{m} e^{x} \left(\frac{d}{dx}\right)^{k-m} (1-x)^{\lambda} \qquad (k \in P)$$

for $0 < x < 1$ and therefore, since $e^{([\alpha]+2)}(v) = O((1-v)^{\lambda-[\alpha]-2})$,

$$I_2 = O\left(\int_{1/2}^{1-2u} x^{\alpha} \int_{x}^{x+u} (1-v)^{\lambda-[\alpha]-2} dv dx\right) = O(u^{\lambda-[\lambda]});$$

analogously to the estimates following (4.8) there holds $I_3 = O(u^{\lambda-[\lambda]})$ and $I_4 = O(u^{\lambda-[\lambda]})$. Thus, by Lemma 3.20 ($\int_0^{1/2} x^{\alpha} |e^{([\alpha]+1)}(x)| dx < \infty$ is trivially satisfied) one finally has $e \in BV_{\alpha+1}$, proving (4.9).

Now the remaining assertions easily follow by (4.9) and Theorems 4.4 ii), iv), 4.6.

<u>Remark.</u> Of course one could have applied Theorems 4.2 and 3.19 to deduce relation ii) of the above theorem directly, even for $\lambda \geq \alpha \geq 0$. But we preferred the argument presented here because it illustrates how relations of type (4.9) and of type

$$\| T(\rho)f - f \| \leq D \| S(\rho)f - f \|$$

(and conversely) make an easy deduction of some approximation results possible.

Let us mention that for $\Phi(k) = k^{\kappa}$, $\kappa > 0$, (as well as for some other particular examples) statement ii) is already proved in [59;I] provided $\lambda \geq \alpha \in P$, iii) in [33;II] provided $\lambda \geq \alpha \in P$, v) in [33;I] provided $\lambda = \alpha = 1$.

4.4 Bessel potentials

Let us introduce the Bessel potentials $\mathcal{L}_{\Phi,\beta}$ defined for $\beta > 0$, $\rho > 0$ by

$$L_{\Phi,\beta}(\rho)f \sim \sum_{k=0}^{\infty}(1 + \Phi(k)/\Phi(\rho))^{-\beta}\, P_k f \qquad (f \in X), \tag{4.10}$$

where Φ is supposed to satisfy the conditions of Lemma 3.8.

These means coincide with the standard Bessel potentials in case $\Phi(k) = k^2$, with the Picard means in case $\Phi(k) = k$. Since $(1+x)^{-\beta} \in BV_{j+1}$ for each $j \in \mathbb{P}$, these means are always defined on X provided (3.7) holds for some $\alpha \geq 0$ in view of Lemma 3.15 ii) and Theorem 3.22.

Since $(1+x)^{\beta}\, e^{-x} \in BV_{j+1}$ for each $j \in \mathbb{P}$, one has

$$\| B^{\psi} W_{\Phi}(\rho) f \| \leq D_1 \| B^{\psi} L_{\Phi,\beta}(\rho) f \| \qquad (f \in X) \tag{4.11}$$

for each $\beta > 0$ and $\psi_k = (\Phi(k))^{\gamma}$, $\gamma > 0$. But the converse is not true. For assume to be so, then by Theorem 4.4 ii) $B^{\psi}L_{\Phi,\beta}(\rho)$ is a continuous operator for each fixed $\rho > 0$ and independent of β,γ with associated multiplier sequence $(\Phi(k))^{\gamma}/(1+\Phi(k)/\Phi(\rho))^{\beta}$, which clearly tends to infinity with k provided $\gamma > \beta$. But this is a contradiction to the continuity of $B^{\psi}L_{\Phi,\beta}$, so that the converse of (4.11) does not hold.

Theorem 4.8. *Let X, $\{P_k\}$ satisfy condition (C^{α}) for some $\alpha \geq 0$; let Φ be given as in Theorem 4.2, $\mathcal{L}_{\Phi,\beta}$, $\beta > 0$, by (4.10) and $\mathfrak{M}_{\Phi}$ by (4.3). Then*

$$\| L_{\Phi,\beta}(\rho) f - f \| \approx \| W_{\Phi}(\rho) f - f \| \qquad (f \in X). \tag{4.12}$$

Furthermore, setting $\psi_k = (\Phi(k))^{\gamma}$ one has

i) $$\| L_{\Phi,\beta}(\rho) f - f \| \leq D_1 (\Phi(\rho))^{-\gamma} |f|_{\psi} \qquad (f \in X^{\psi})$$

for $0 < \gamma \leq 1$;

ii) $$\| B^{\psi} L_{\Phi,\beta}(\rho) f \| \leq D_2 (\Phi(\rho))^{\gamma} \| f \| \qquad (f \in X)$$

for $0 < \gamma \leq \beta$;

iii) a) $$\Phi(\rho) \| L_{\Phi,\beta}(\rho) f - f \| = o(1) \qquad (\rho \to \infty)$$

implies $f \in P_o(X)$ and $L_{\Phi,\beta}(\rho) f = f$ for all $\rho > 0$;

b) the Favard class of $\mathcal{L}_{\Phi,\beta}$ *is the set* $(X^{\psi})^{\sim X}$ *where* $\psi_k = \Phi(k)$, *thus* $\gamma = 1$, *and*

$$\sup_{\rho>0} \Phi(\rho)\|L_{\Phi,\beta}(\rho)f - f\| \sim |f|_{\psi^\sim}$$

$$\sim \sup_{n\in\mathbb{N}} \|\textstyle\sum_{k=0}^{n} (A_{n-k}^{\alpha}/A_n^{\alpha})\Phi(k)P_k f\|$$

are equivalent semi-norms on $(X^{\psi})^{\sim X}$;

iv)
$$\|B^{\psi}L_{\Phi,\beta}(\rho)f\| \leq D_3(\Phi(\rho))^{\gamma}\|L_{\Phi,\beta}(\rho)f - f\| \qquad (f \in X)$$

provided $\beta \geq 1$, $1 \leq \gamma \leq \beta$;

v)
$$\|L_{[\Phi]^{\gamma},\beta}(\rho)f-f\| \leq D_4 \|L_{\Phi,\beta}(\rho)f-f\| \qquad (f \in X)$$

for all $\gamma > 1$.

Proof. For the validity of (4.12) it is sufficient to show by Theorem 3.9 and Lemma 3.15 ii) that $\{e(x)-1\},\{[e(x)]^{-1} - 1\} \in BV_{j+1}$ for some $j \geq \alpha$, $j \in P$, where

$$e(x) = \frac{1-(1+x)^{-\beta}}{1-e^{-x}}, \quad [e(x)]^{-1} = \frac{1-e^{-x}}{1-(1+x)^{-\beta}} .$$

Differentiating $e(x)$

$$e'(x) = \frac{\beta(1+x)^{-\beta-1}}{1-e^{-x}} - \frac{(1-(1+x)^{-\beta})e^{-x}}{(1-e^{-x})^2} \equiv Z(x)/(1-e^{-x})^2$$

one readily verifies that $Z^{(k)}(x) = O(x^{2-k})$ for $x \to 0+$ and hence, analogously to the proof of Theorem 4.4, it follows that $\int_0^1 x^j |e^{(j+1)}(x)|dx < \infty$ for each (fixed) $j \in P$. A further differentiation of $e'(x)$ yields $e^{(j+1)}(x) = O(x^{-\beta-j-2})$ for $x \to \infty$, and hence $e \in BV_{j+1}$ for each fixed $j \in P$. An analogous argumentation also delivers $[e(x)]^{-1} \in BV_{j+1}$, and thus (4.12) is established.

Now, Theorem 4.4 combined with relation (4.12) immediately proves i), iii), v); hence, only assertions ii) and iv) have still to be discussed. Again, by Theorem 3.9 and Lemma 3.15 ii), one has to demonstrate that $e_1(x), e_2(x) \in BV_{j+1}$, $j \in P$, where

$$e_1(x) = x^{\gamma}(1+x)^{-\beta}, \quad e_2(x) = x^{\gamma}(1+x)^{-\beta}/(1-(1+x)^{-\beta}) .$$

An elementary discussion gives $e_1^{(j+1)}(x) = O(x^{\gamma-1-j})$ for $x \to 0+$, $0 < \gamma \leq \beta$, and

$$e_1^{(j+1)}(x) = O\left(\begin{cases} x^{\gamma-\beta-1-j} & , 0 < \gamma < \beta \\ x^{-j-2} & , \gamma = \beta \end{cases}\right) \qquad (x \to \infty).$$

Hence $e_1 \in BV_{j+1}$. Analogously, setting $e_2'(x) = Z(x)/((1+x)^{\beta}-1)^2$, one obtains $Z^{(k)}(x) = O(x^{\gamma-k})$ for $x \to 0+$, $\gamma > 1$, and $Z^{(k)}(x) = O(x^{2-k})$ for $x \to 0+$ provided $\gamma = 1$; hence $\int_0^1 x^j |e_2^{(j+1)}(x)| dx < \infty$, as in the proof to Theorem 4.4. Since furthermore

$$e_2^{(j+1)}(x) = O\left(\begin{cases} x^{\gamma-\beta-j-1}, & 1 \leq \gamma < \beta \\ x^{-2-j} & , \gamma = \beta \geq 1 \end{cases}\right) \qquad (x \to \infty),$$

one has $e_2 \in BV_{j+1}$, so that Theorem 4.8 is completely proved (in case $\gamma = \beta$ cf. Footnote 8).

For $\Phi(k) = k^2$, $0 < \gamma < \beta$, statement ii) is already proved in [59;I]. In that paper there is indicated another generalization which, in our framework, reduces to a discussion of $\psi(x)(1+x)^{-\beta}$, where $\psi(0) = 0$ and $\lim_{x\to\infty} \psi(x) = \infty$; essentially it is a priori assumed there that $\psi(x)(1+x)^{-\beta}$ belongs to BV_{j+1}.

4.5 Cesàro means

The Cesàro means of order β (i.e., the (C,β)-means) have been introduced in Sec. 3.1 by

$$(4.13) \qquad (C,\beta)_n f = \sum_{k=0}^{n} \{ A_{n-k}^{\beta}/A_n^{\beta} \} P_k f \qquad (f \in X).$$

Since $(C,\beta)_n$ is a polynomial summation method, Theorem 4.2 and (3.8) immediately yield the Bernstein-type inequality

$$(4.14) \quad \| \sum_{k=0}^{n} \{A_{n-k}^{\beta}/A_n^{\beta}\} \Phi(k) P_k f \| \leq D^* \Phi(n) \| (C,\beta)_n f \| \leq D\Phi(n) \| f \|$$

for all $\beta \geq \alpha \geq 0$ (D being independent of n and f,Φ satisfying the hypotheses of Theorem 3.9).

For an application of the general approximation Theorem 1.2 it only remains to derive a suitable Jackson-type inequality. We will settle this problem by solving the saturation problem. In order to avoid the verification of the multiplier condition (2.11), we make use of the functional equation of the Cesàro means, thus following Berens-Butzer-Pawelke [16] and others (who applied this method to concrete orthogonal expansions).

Theorem 4.9. *Let X, $\{P_k\}$ satisfy condition (C^α) for some $\alpha \geq 0$. Then, for each $\beta \geq \alpha$, $\beta > 0$, the (C,β)-means have the following properties:*

a) $$\|(C,\beta)_n f - f\| = o(n^{-1}) \qquad (n \to \infty)$$

implies $f \in P_o(X)$ and $(C,\beta)_n f = f$ for all $n \in \mathbf{P}$;

b) the Favard class of $(C,\beta)_n$ is the set $(X^\varphi)^{\sim X}$ with $\varphi_k = \beta k$ and the following semi-norms are equivalent on $(X^\varphi)^{\sim X}$:

i) $\sup_{n \in \mathbf{P}} n\|(C,\beta)_n f - f\|$, *ii)* $|f|_{\varphi^\sim}$,

iii) $\sup_{n \in \mathbf{P}} \|\sum_{k=o}^{n} (A_{n-k}^{\beta}/A_n^{\beta}) k\, P_k f\|$.

Proof. Since (cf. [27;p.388])

$$(4.15) \qquad \lim_{n\to\infty} n\{1 - A_{n-k}^{\beta}/A_n^{\beta}\} = \beta k \qquad (k \in \mathbf{P}),$$

part a) follows by Lemma 2.3 . Furthermore, the equivalence of ii) and iii) is given by Theorem 2.6 a), and the implication i) $\Longrightarrow$ iii) follows by Theorem 3.1 on account of (4.15). Thus it remains to show that iii) implies i).

Now on account of the identity [110;I, p.269]

$$(C,\beta+1)_n f - (C,\beta+1)_{n-1} f = \frac{\beta+1}{n(n+\beta+1)} \sum_{k=o}^{n} (A_{n-k}^{\beta} / A_n^{\beta}) k\, P_k f$$

it follows for $m > n$ that

$$(C,\beta+1)_m f - (C,\beta+1)_n f = \sum_{l=n+1}^{m} \frac{\beta+1}{l(l+\beta+1)} \sum_{k=o}^{l} (A_{n-k}^{\beta}/A_n^{\beta}) k\, P_k f .$$

Since the sequence $\{P_k\}$ is fundamental and all (C,β)-means, $\beta \geq \alpha$, are bounded (see (3.8)), one has by the Banach-Steinhaus theorem that $(C, \beta+1)_m f$ tends to f for $m \to \infty$. Thus,

$$\| f - (C,\beta+1)_n f\| \leq \sum_{l=n+1}^{\infty} \frac{\beta + 1}{l(l+\beta+1)} \sup_{l\in N} \| \sum_{k=o}^{l} (A_{n-k}^{\beta}/A_n^{\beta})kP_k f\| .$$

Now the first sum is bounded by $(\beta+1)n^{-1}$, and iii)$\Longrightarrow$i) is proved with β replaced by $\beta + 1$. With the following identity it is easy to proceed from $\beta + 1$ to β. For, on account of

$$(C,\beta)_n f = (C,\beta+1)_n f + \frac{1}{n+\beta+1} \sum_{k=o}^{n} (A_{n-k}^{\beta}/A_n^{\beta})kP_k f$$

one finally has for all $\beta \geq \alpha$

$$n\| f-(C,\beta)_n f\| \leq n\| f-(C,\beta+1)_n f\| + \frac{n}{n+\beta+1}\| \sum_{k=o}^{n}(A_{n-k}^{\beta}/A_n^{\beta})kP_k f \|$$

$$\leq (\beta+2) \sup_{n\in N}\| \sum_{k=o}^{n} (A_{n-k}^{\beta}/A_n^{\beta})kP_k f\| .$$

Theorem 4.9 has essentially been proved by Alexits [1] for $\alpha = \beta = 1$, by Favard [43] for $\alpha = \beta \in N$, and by Li Sjun-Czin [69] for $\alpha = \beta > 0$ (one may suppose $\alpha = \beta$ without loss of generality by (3.8)); for $\beta \geq \alpha \in N$ Theorem 4.9 exactly coincides with that in [33;II]. Let us mention that [1], [69] treat Banach valued series $\sum f_k$, the elements f_k not necessarily being orthogonal, and that [69] develops several extensions .

4.6 de La Vallée-Poussin means

Let us introduce the summation method of de La Vallee - Poussin 10) by ($n \in P$)

$$(4.16) \qquad V_n f = \sum_{k=o}^{n} \omega_k(n)P_k f,\ \omega_k(n) = \begin{cases} \frac{(n!)^2}{(n-k)!(n+k)!}, & 0 \leq k \leq n \\ 0, & k > n . \end{cases}$$

V_n is a polynomial operator on $(X, \{P_k\})$, and hence Theorem 4.2 delivers

$$\| B^{\psi} V_n f \| \leq D^* \Phi(n) \| V_n f \| \qquad (f \in X), \tag{4.17}$$

where $\psi_k = \Phi(k)$ and Φ satisfies the hypotheses of Theorem 3.9. To obtain the analog of (4.14) one has to estimate $\| V_n \|$. The uniform boundedness of $\| V_n \|$ in case of numerical series, provided (3.7) holds for some $\alpha \geq 0$, has been established in several papers (see [108;p.154]). The proof by Rychlêk [88] carries over to our situation without any difficulties and it is sketched here briefly. Basic is the following lemma (quoted without proof) due to Rychlêk [88].

Lemma 4.10. *a) For sufficiently large n and arbitrary $j \in \mathbb{P}$ one has*

$$|\Delta^{j+1} \omega_k(n)| \leq D\, n^{-(j+1)/2} \qquad (0 \leq k \leq n),$$

the constant being independent of n.

b) For $m = [(1 + \frac{j+1}{\sqrt{2}}) \sqrt{n}]$ and $j \in \mathbb{P}$ one has $\Delta^{j+1} \omega_k(n) \geq 0$, $m \leq k \leq n$.

As Rychlêk [88] has shown it is easy to verify that $\{\omega(n)\}_{n \in \mathbb{P}} \subset bv_{j+1}$ uniformly in n for each $j \in \mathbb{P}$. Indeed, with m as in Lemma 4.10 b), consider

$$\Big(\sum_{k=0}^{m-1} + \sum_{k=m}^{n}\Big) A_k^j |\Delta^{j+1} \omega_k(n)| \equiv S_1 + S_2 .$$

Then by Lemma 4.10 a), relations (3.5) and (3.6) it follows that

$$S_1 \leq D\, n^{-(j+1)/2} \sum_{k=0}^{m-1} A_k^j = D\, n^{-(j+1)/2} A_{m-1}^{j+1} \leq D_1 .$$

On account of Lemma 4.10 b) one obtains for S_2

$$S_2 = \sum_{k=m}^{n} A_k^j \Delta^{j+1} \omega_k(n) = A_{m-1}^{j} \Delta^j \omega_m(n) + \sum_{k=m}^{n} A_k^{j-1} \Delta^j \omega_k(n) .$$

A repeated application of this argument leads to

$$S_2 = \sum_{k=0}^{j} A_{m-1}^{k} \Delta^k \omega_m(n) \leq D^* \sum_{k=0}^{j} (m-1)^k n^{-k/2} \leq D_2$$

by Lemma 4.10 a) and relation (3.6). Since $m \sim \sqrt{n}$ and the sum has only (j+1) terms, D_2 is independent of n. Thus, we may estimate (4.17) further, obtaining

$$(4.17)' \qquad \| B^{\psi} V_n f \| \leq D\, \Phi(n) \| f \| \qquad (f \in X)$$

with $\psi_k = \Phi(k)$ and Φ satisfying the hypotheses of Theorem 3.9. Now, on account of the asymptotic relation (see e.g. [92;p.68])

$$(4.18) \qquad \omega_k(n) = e^{-k^2/n}\{1 + O(n^{-2})\},$$

it is obvious that $\lim_{n\to\infty} \omega_k(n) = 1$ for each $k \in \mathbb{P}$, and thus $\mathcal{W}$ is a summation method by the Banach-Steinhaus theorem. To derive a Jackson-type inequality we again solve the saturation problem and, analogously to Butzer-Pawelke [34] (cf. [31;p.448]), make use of the functional equation typical for the de La Vallée-Poussin means, namely,

$$(4.19) \qquad k^2 \omega_k(n) = n^2\{\omega_k(n) - \omega_k(n-1)\} \qquad (n \in \mathbb{N}, k \in \mathbb{P}).$$

In particular we show

Theorem 4.11. *Let X, $\{P_k\}$ satisfy condition (C^{α}) for some $\alpha \geq 0$; let $\mathcal{W}$ be given by (4.16).*

a)
$$\| V_n f - f \| = o(n^{-1}) \qquad (n \to \infty)$$

implies $f \in P_o(X)$ and $V_n f = f$ for all $n \in \mathbb{P}$;

b) the Favard class of $\mathcal{W}$ is the set $(X^{\varphi})^{\sim X}$ with $\varphi_k = k^2$, and the following semi-norms are equivalent on $(X^{\varphi})^{\sim X}$:

$$i) \quad \sup_{n \in \mathbb{P}} n \| V_n f - f \|, \qquad ii) \quad |f|_{\varphi^{\sim}}$$

$$iii) \quad \sup_{n \in \mathbb{P}} \| \textstyle\sum_{k=o}^{n} (A_{n-k}^{\alpha}/A_n^{\alpha}) k^2 P_k f \|.$$

Proof. By Lemma 2.3 and (4.18) assertion a) follows since

$$(4.20) \qquad \lim_{n\to\infty} n\{1 - \omega_k(n)\} = k^2 .$$

The equivalence of ii) and iii) in b) is given by Theorem 2.6 a), the implication i) $\Longrightarrow$ iii) by Theorem 3.1 on account of (4.20). Thus it remains to show that iii) implies i).

Now, using the functional equation (4.19) and proceeding as in the proof to Theorem 4.9, one obtains

$$V_m f - V_n f = \sum_{l=n+1}^{m} \frac{1}{l^2} B^{\varphi} V_l f$$

and passing to the limit for $m \to \infty$ ($\mathfrak{W}$ being a summation method),

$$\| V_n f - f \| \leq \sum_{l=n+1}^{\infty} l^{-2} \sup_{l \in \mathbb{N}} \| B^{\varphi} V_l f \| \leq n^{-1} \sup_{k \in \mathbb{P}} \| B^{\varphi} V_k f \| .$$

Now $(C,\alpha)_n$ and V_n map X into the domain of B^{φ} and therefore, by Theorem 2.6 a),

$$\sup_{n \in \mathbb{P}} \| \sum_{k=o}^{n} (A_{n-k}^{\alpha} / A_n^{\alpha}) k^2 \, P_k f \| \sim \sup_{n \in \mathbb{P}} \| B^{\varphi} V_n f \|$$

and the theorem is established completely.

Remark. Since we have not verified the multiplier condition (2.11) for the last two examples, i.e., for the Cesàro and de La Vallée-Poussin means, we cannot immediately conclude that Voronovskaja relations of type (2.13) hold for these two processes. However, a closer examination of the functional equation method shows that a slight modification also gives Voronovskaja type relations for the Cesàro and de La Vallée-Poussin means, as given e.g. in [16] and [31; p. 449], respectively; for the general "functional equation approach" see Butzer-Pawelke [34].

Choosing in (4.17)' $\psi_k = k^2$ we see that $\| B^{\psi} V_n f \| \leq D\, n^2 \| f \|$, whereas the saturation order given in Theorem 4.11 is only n^{-1}; furthermore, it is known in the trigonometric series case (see Butzer-Scherer [35; p.138]) that the order n^2 in the above Bernstein type inequality may be replaced by n (for Jacobi polynomials see [11; p. 60]). Both proofs essentially use the closed representation of the kernel (associated with de La Vallée-Poussin's method) in the concrete setting of the trigonometric system (or Jacobi polynomials). Here we will give a proof of $\| B^{\psi} V_n f \| \leq D n \| f \|$ depending only upon the properties of the coefficients $\omega_k(n)$ (cf. (4.16)). Indeed

Theorem 4.12. Let X, $\{P_k\}$ satisfy condition (C^{α}) for some $\alpha \geq 0$; let $\mathfrak{W}$ be given by (4.16). Then for $\{\psi_k\} = \{k^2\}$ one has

$$\| B^{\psi} V_n f \| \leq D n \| f \| \qquad (f \in X), \tag{4.21}$$

the constant D being independent of f and n.

Proof. By Theorem 3.3 and Lemma 3.2 it is sufficient to show that $\{(k^2/n)\omega_k(n)\}_{k=0}^{\infty}$ belongs to bv_{j+1} for some $j \geq \alpha$, $j \in P$, uniformly in $n \in N$. First we remark that for arbitrary $\eta \in s$ there holds the identity

$$(4.22) \qquad \Delta^{j+1}\, \eta_k \omega_k(n) = \sum_{i=0}^{j+1} \binom{j+1}{i} \Delta^m \eta_k \; \Delta^{j+1-i} \omega_{k+i}(n),$$

and therefore, with $\eta_k = k^2/n$,

$$\Delta^{j+1}(k^2/n)\omega_k(n) = (k^2/n)\,\Delta^{j+1}\omega_k(n) - (j+1)\{(2k+1)/n\}\,\Delta^j\omega_k(n)$$

$$+ \{(j+1)j/n\}\,\Delta^{j-1}\omega_{k+2}(n) \quad .$$

Now proceed as in [88] : choose m as in Lemma 4.10 and consider

$$\left(\sum_{k=0}^{m-1} + \sum_{k=m}^{n}\right) A_k^j \,|\Delta^{j+1}(k^2/n)\omega_k(n)| \equiv S_1 + S_2 \ .$$

Using again Lemma 4.10 a), relations (3.5) and (3.6), it follows readily that ($m \sim \sqrt{n}$)

$$S_1 \leq \sum_{k=0}^{m-1} A_k^j (k^2/n)\,|\Delta^{j+1}\omega_k(n)| + \sum_{k=0}^{m-1} A_k^j \{(j+1)(2k+1)/n\}\,|\Delta^j\omega_k(n)|$$

$$+ \{(j+1)j/n\} \sum_{k=0}^{m-1} A_k^j \,|\Delta^{j-1}\omega_{k+2}(n)| < \infty$$

uniformly in n. To estimate S_2 make use of the positivity of the differences (see Lemma 4.10 b)) and reduce the order of the differences. Thus it follows first that

$$S_2 \leq \sum_{k=m}^{n}(k^2/n)A_k^{j-1}\Delta^j\omega_k(n) + \{(j+2)/n\}\sum_{k=m}^{n} A_{k-1}^j(2k-1)\Delta^j\omega_k(n)$$

$$+ \{j(j+1)/n\}\sum_{k=m}^{n} A_{k-2}^j \;\Delta^{j-1}\omega_k(n)$$

$$+ A_{m-1}^j\{(m-1)^2/n\}\Delta^j\omega_m(n) - A_{m-1}^j\{(j+1)(2m-1)/n\}\;\Delta^j\omega_m(n)$$

$$- \{j(j+1)/n\}A_{m-2}^j\;\Delta^{j-1}\omega_m(n) - \{j(j+1)/n\}A_{m-1}^j\;\Delta^{j-1}\omega_{m+1}(n) \ .$$

The last four terms are uniformly bounded in n by (3.6) and Lemma 4.10 a). Thus, repeating this procedure leads to

$$S_2 < \sum_{k=m}^{n} (k^2/n)A_k^{j-2}\,\Delta^{j-1}\omega_k(n) + \{(j+3)/n\}\sum_{k=m}^{n} A_{k-1}^{j-1}(2k-1)\Delta^{j-1}\omega_k(n)$$

$$+ \{(j(j+1) + 2(j+2))/n\}\sum_{k=m}^{n} A_{k-2}^{j}\,\Delta^{j-1}\omega_k(n) + O(1) \equiv \sum\nolimits_1 + \sum\nolimits_2 + \sum\nolimits_3 + O(1),$$

the single terms arising anew being again uniformly bounded in n. Now reducing the order of the differences from (j-1) to (j-2) effects three single terms uniformly bounded in n on the one hand, and on the other, $\sum_1$ and $\sum_2$ produce contributions to $\sum_2$ and $\sum_3$, respectively. Since j is finite, a (j-2)-times iteration leads to

$$S_2 = O(1) + \sum_{k=m}^{n}(k^2/n)\Delta\omega_k(n)$$

$$+ O(n^{-1}\sum_{k=m}^{n}(2k-1)A_{k-1}^{1}\Delta\omega_k(n) + n^{-1}\sum_{k=m}^{n} A_{k-2}^{2}\Delta\omega_k(n))$$

$$= O(1) + O(\sum_{k=m}^{n} (k/n)\omega_k(n)).$$

Using the asymptotic relation (4.18) it immediately follows that

$$\sum_{k=m}^{n}(k/n)\omega_k(n) = O(\int_{\sqrt{n}}^{n} \frac{x}{n}\, e^{-x^2/n}\{1 + O(n^{-2})\}dx) = O(1)$$

is uniformly bounded in n, and the assertion is proved.

<u>Remark.</u> Replacing k^2/n by $k/\sqrt{n}$, a slight modification of the above argument yields a relation analogous to (4.21). More generally, one may replace k^2/n by $k^l/n^{l/2}$, $l \in \mathbb{N}$, a fact, we will indicate briefly.

By (4.22) and $\Delta^{l+i}\,k^l = 0$ for $i \in \mathbb{N}$ one has ($l^* = \min\{l,j+1\}$)

$$\Delta^{j+1}(k^l/n^{l/2})\omega_k(n) = \sum_{i=0}^{l^*}\binom{j+1}{i}\Delta^i(k^l/n^{l/2})\Delta^{j+1-i}\omega_{k+i}(n) .$$

Proceeding as above gives $S_1 = O(1)$ uniformly in n, noting that $\Delta^i\,k^l \sim k^{l-i}$. S_2 may be estimated by a finite linear combination (depending upon l) of terms of the following type

$$(4.23)\quad \sum\nolimits_i = \sum_{k=m}^{n} A_{k-i}^{j}|n^{-l/2}\Delta^i(k-i)^l|\Delta^{j+1-i}\omega_k(n)$$

$$- \sum_{k=m}^{m+i-1} A_{k-i}^{j}|n^{-l/2}\Delta^i(k-i)^l|\Delta^{j+1-i}\omega_k(n),$$

where the second sum is uniformly bounded in n since $0 \leq i \leq l$. A reduction repeated $(j + 1 - i)$-fold shows that S_2 may be estimated by $O(1)$ (uniformly bounded in n, depending upon l and j) and finite linear combination of

$$\sum_i^* = \sum_{k=m}^n A_{k-i}^{i-1} \, |n^{-1/2}\Delta^i(k-i)^l| \, \omega_k(n) \qquad (0 \leq i \leq l^*).$$

Let $n \geq 4$ without loss of generality; then $\sum_o^* = 0$ since $A_k^{-1} = 0$ for $k \geq 1$. By (4.18), (3.6), and $\Delta^i(k-i)^l \sim k^{l-i}$ it follows that

$$\sum_i^* = O(\sum_{k=m}^n k^{i-1} \, k^{l-i} \, n^{-1/2}\omega_k(n))$$
$$= O(\int_{\sqrt{n}}^n n^{-1/2} \, x^{l-1} \, e^{-x^2/n} \, dx) = O(1)$$

is uniformly bounded in n. Since l,j are finite numbers there is only a finite number (not depending on n) of terms which are uniformly bounded in n. Thus, we have proved

$$(4.24) \qquad \| \sum_{k=o}^n k^l \frac{(n!)^2}{(n-k)!\,(n+k)!} P_k f\| \leq D \, n^{l/2} \| f\| \qquad (l \in \mathbb{N})$$

A refinement of this method should also deliver appropriate Bernstein-type inequalities with respect to $\{k^\beta\}$, $\beta > 0$, $\{\log(1+k)\}$, and some other particular examples. A direct application of Theorem 3.9 seems to be hard.

5. APPLICATIONS TO PARTICULAR EXPANSIONS

In this section we consider concrete choices of the Banach space X together with its sequence of projections. In Subsection 5.1 the standard one-dimensional trigonometric system is treated, in 5.2 its n-dimensional version. In 5.3 expansions into Laguerre and Hermite series are considered, in 5.4 expansions into Jacobi polynomials; we conclude with a discussion of expansions into spherical harmonics in 5.5. It is known that all these expansions satisfy the basic hypothesis (3.7), namely that the corresponding (C,α)-means are uniformly bounded for certain values $\alpha \geq 0$. Of course there are other expansions also satisfying this property, e.g. expansions into Bessel functions (see Wing [107], Benedek-Panzone [12,13,14]), into more general eigenfunctions (of appropriate initial value problems) (see Turner [105]), into Walsh and Haarfunctions, etc.; for a discussion of these series (with further comments and literature) we refer to [59,II].

Let us mention that i) the following applications to n-dimensional Fourier series as well as to Jacobi series are new with respect to [33;I,II], [59;I,II]; ii) parts of our applications for integer α in 5.1, 5.3, and 5.5 coincide with those in [33;I,II], [59;I,II], and we will not refer to the latter papers if the results are standard.

At this point the author would like to thank Professors R. Askey, B. Muckenhoupt, and P. Sjölin for several written communications. R. Askey provided us with references concerning applications to Jacobi series and hinted at the "transplantation" approach (see Sec. 1.1) as an alternative method of proving some of the present results. B. Muckenhoupt informed us of the newest results on (C,α)-boundedness of Laguerre and Hermite series, and P. Sjölin communicated the last refinements on Bochner-Riesz summability of multidimensional Fourier series.

5.1 One-dimensional trigonometric system

Let $X_{2\pi} = L^p_{2\pi}$, $1 \leq p < \infty$, or $C_{2\pi}$ be the Banach space of 2π-periodic functions with standard norms $\|\cdot\|_{X_{2\pi}}$

$$\{ \int_{-\pi}^{\pi} |f(x)|^p dx\}^{1/p} \quad (1 \leq p < \infty) \text{ or } \max_x |f(x)|,$$

respectively. Defining $\{P_k\}$ by

$$(5.1) \qquad P_0 f(x) = f^\wedge(0), \; P_k f(x) = f^\wedge(k)e^{ikx} + f^\wedge(-k)e^{-ikx} \quad (k \in \mathbb{N}),$$

$f^\wedge(k)$ being the usual Fourier coefficients

$$f^\wedge(k) = \frac{1}{2\pi}\int_{-\pi}^{\pi} f(x)e^{-ikx}dx \qquad (k \in \mathbb{Z}),$$

it is obvious that $\{P_k\}$ is a sequence of orthogonal projections which is total on account of the uniqueness property for Fourier series and fundamental, since the linear span of $\bigcup_{k\in \mathbb{Z}} e^{ikx}$ is dense in $X_{2\pi}$ (however not in $L^\infty_{2\pi}$; density is only needed for convergence statements; cf. Footnote 1).

Concerning the fundamental hypothesis (3.7) the famous theorem of M. Riesz states that the partial sums $(C,0)_n = \sum_{k=0}^{n} P_k f$ of the Fourier series are uniformly bounded in n provided $1 < p < \infty$, but not for $p = 1$ and $p = \infty$. Furthermore, it has long been known that the operators $(C,\alpha)_n$ are also uniformly bounded in n for each $\alpha > 0$ on all $X_{2\pi}$ spaces (see e.g. Chapman [40]). Thus we may apply the results of Sec. 4, and we begin with a Bernstein-type inequality (Theorem 4.2).

Corollary 5.1. *Let $X_{2\pi}$ and $\{P_k\}$ be given as above. Choose $\Phi(x) = x^\beta \log^\gamma(1+x^\delta)$ with $\beta, \gamma, \delta > 0$ or $\beta = 0$, $\gamma,\delta > 0$ or $\beta > 0$, $\gamma = 0$. Then, for arbitrary $c_k \in \mathbb{C}$ $(k \in \mathbb{Z})$,*

$$\left\| \sum_{k=-n}^{n} |k|^\beta \log^\gamma(1+|k|^\delta) c_k e^{ikx} \right\| \leq D\, n^\beta \log^\gamma(1+n^\delta) \left\| \sum_{k=-n}^{n} c_k e^{ikx} \right\|,$$

the constant D being independent of n and $\{c_k\}$.

For $\gamma = 0$ this result is standard (see e.g. [57, 58]); for $\beta = 0$, $\gamma = 1$ and $\delta = 1$ or 2 this is a result of Görlich [57,58] who shows in the trigonometric series case that for all (even or uneven) convex or concave sequences $\{\Phi(k)\}_{k \in P}$ with $\Phi(0) = 0$ and $\lim_{k\to\infty}\Phi(k) = \infty$ our constant D (which naturally depends upon C_α (in (3.7)), β, γ, δ and may e.g. tend to infinity for $\beta \to \infty$) can be replaced by 2; in particular, Görlich proved that the multiplier sequence $\{e^k\}$, restricted to trigonometric polynomials of degree n, has norm less than or equal to $2\,e^n$, whereas our calculus would only yield the bound $D_\alpha n^\alpha e^n$ ($\alpha > 0$ arbitrary small).

Comparing Corollary 5.1 with the classical Bernstein inequality

$$(5.2) \qquad \| \frac{d}{dx} \sum_{k=-n}^{n} c_k e^{ikx} \| \leq D\, n \| \sum_{k=-n}^{n} c_k e^{ikx} \| ,$$

one is surprised that the latter inequality is not included. However, by using the theory of conjugate functions (cf. [31;Ch.9]) one may derive (5.2) provided $1 < p < \infty$. Indeed, observing that

$$(ik)^{\beta} = |k|^{\beta} (\cos\tfrac{\pi}{2}\beta + i \operatorname{sgn} k \sin\tfrac{\pi}{2}\beta)$$

and

$$\| \sum_{k=-n}^{n} i \operatorname{sgn} k\, c_k e^{ikx} \|_p \leq A_p \| \sum_{k=-n}^{n} c_k e^{ikx} \|_p \qquad (1 < p < \infty),$$

one immediately has for $\beta > 0$

$$\| \sum_{k=-n}^{n} (ik)^{\beta} c_k e^{ikx} \|_p \leq D^* n^{\beta} \| \sum_{k=-n}^{n} c_k e^{ikx} \|_p \qquad (1 < p < \infty),$$

and thus in particular the above inequality (5.2) for $\beta = 1$. Since apart from this special argument the present general approach only admits even two-a-way sequences $\{\Phi(k)\}_{k \in Z}$, there seem to be principal difficulties in $L^1_{2\pi}$ or $C_{2\pi}$.

Concluding the discussion of Bernstein-type inequalities we apply (4.24)

Corollary 5.2. Let $X_{2\pi}$ and $\{P_k\}$ be given as above. Then for $\beta \in N$

$$\| \sum_{k=-n}^{n} |k|^{\beta} \frac{(n!)^2}{(n-|k|)!(n+|k|)!} f^{\wedge}(k) e^{ikx} \| \leq D\, n^{\beta/2} \| f \| \qquad (f \in X),$$

D being independent of n and f.

For $\beta = 2$ this corollary coincides with a result of Butzer-Scherer [35;p.138], whereas other β values do not seem to have been considered.

Concerning the saturation problem for the trigonometric system we first of all remark that on account of the close interaction between periodic and non-periodic Fourier multipliers (see e.g. [97;Ch.7]) all saturation results on the real line are also valid for the circle; furthermore, beginning with Butzer [24], Harsiladze [62], and Sunouchi-Watari [99], very many authors dealt with this problem. Therefore we only mention three longer articles of Butzer-Görlich [27], Sunouchi [98], and Tureckii [104] and generally refer to the book of Butzer-

Nessel [31; Ch. 12-13] for many results and a survey of the literature.

Now choosing e.g. $\Phi(x) = x^\beta$, $\beta > 0$, Theorem 4.4 iii) is contained in [31;p.447], Theorem 4.7 iii) in [31;p.446, p.475], Theorem 4.8 iii), $\beta = 1$, in [31;p.465], Theorem 4.9 in [27], Theorem 4.11 in [31;p.449]. Most of the known results correspond to the special choice $\Phi(x) = x^\beta$, whereas we still have a number of possibilities in choosing Φ. Thus e.g. Theorem 4.4 iii) yields

<u>*Corollary 5.3.*</u> *Let $X_{2\pi}$ and $\{P_k\}$ be given as above and choose $\Phi(x) = x^\beta \log(1+x^\delta)$, $\beta, \delta > 0$. Then*

a) $$\Big\| \sum_{k=-\infty}^{\infty} (1+|k|^\delta)^{-\varepsilon|k|^\beta} f^\wedge(k) e^{ikx} - f(x) \Big\| = o(\varepsilon) \qquad (\varepsilon \to 0+)$$

implies that f is a constant;

b) $$\Big\| \sum_{k=-\infty}^{\infty} (1+|k|^\delta)^{-\varepsilon|k|^\beta} f^\wedge(k) e^{ikx} - f(x) \Big\| = O(\varepsilon) \qquad (\varepsilon \to 0+)$$

if and only if

$$\Big\| \sum_{k=-n}^{n} \Big(1 - \frac{|k|}{n+1}\Big) |k|^\beta \log(1+|k|^\delta) f^\wedge(k) e^{ikx} \Big\| = O(1) \qquad (n \to \infty).$$

To give an example of a Zamansky-type inequality we apply Theorem 4.7 iv) with $\Phi(x) = x^\beta$ and write $R_{\Phi,\lambda}(\rho) = R_{\beta,\lambda}(\rho)$.

<u>*Corollary 5.4.*</u> *Let $X_{2\pi}$ and $\{P_k\}$ be given as above. Then, for $0 < \beta \leq 2$ and $\lambda > 0$,*

$$\Big\| \frac{d^2}{dx^2} R_{\beta,\lambda}(\rho) f(x) \Big\| \leq D\rho^2 \| R_{\beta,\lambda}(\rho) f - f \|,$$

the constant D being independent of ρ and $f \in X_{2\pi}$.

Finally, arriving at the comparison problem (see Def. 1.4), we mention that in case $\Phi(x) = x^\beta$, $\beta > 0$, and $\lambda = 1$ Theorem 4.6 is contained in [31;p.495,p.501]. In particular, for $\beta = 1$, $\lambda = 1$ Theorem 4.6 implies the equivalence (with respect to rate of growth) of the singular integral of Fejér

$$\sigma_n f(x) = \frac{1}{2\pi(n+1)} \int_{-\pi}^{\pi} f(x-u) \Big[\frac{\sin((n+1)u/2)}{\sin(u/2)} \Big]^2 du$$

with the classical Abel-Poisson integral

$$A_r f(x) = \frac{1}{2\pi} \int_{-\pi}^{\pi} f(x-u) \frac{1 - r^2}{1-2r\cos u + r^2}\, du \qquad (r = e^{-1/n}),$$

a result due to Shapiro [89] and Žuk [109].

5.2 Multiple Fourier series

Let R^n be the n-fold Cartesian product of R and denote by u,v the elements of R^n: $v = (v_1,\ldots,v_n)$; let Z^n be the n-fold Cartesian product of Z with elements $m = (m_1,\ldots,m_n)$. T^n the n-dimensional torus is given by $T^n = \{v \in R^n;\ |v_l| \leq \pi,\ 1 \leq l \leq n\}$.

Choose $X = L^p(T^n)$, $1 \leq p < \infty$, or $C(T^n)$, the set of all functions 2π-periodic in each coordinate with standard norms $\|\cdot\|_X$

$$\{ \int_{T^n} |f(v)|^p dv\}^{1/p} \quad (1 \leq p < \infty) \text{ or } \max_{v \in R^n} |f(v)|,$$

respectively. Now define the projections $\{P_k\}$ by

$$(5.3) \qquad P_k f(v) = \sum_{|m|^2=k} f^\wedge(m) e^{im\cdot v} \qquad (k \in P),$$

where $|m|^2 = \sum_{l=1}^n m_l^2$, $m\cdot v = \sum_{l=1}^n m_l v_l$, and $f^\wedge(m)$ is the standard Fourier coefficient

$$f^\wedge(m) = (2\pi)^{-n} \int_{T^n} f(v) e^{-im\cdot v} dv \qquad (m \in Z^n).$$

(In case there exist no $m \in Z^n$ with $|m|^2 = k$, take e.g. $n = 2$, $k = 3$, it is tacitly assumed that P_k is the null operator which clearly satisfies the orthogonality relation $P_j P_k = \delta_{j,k} P_k$.) Obviously, the projections P_k are mutually orthogonal; since all $e^{im\cdot v}$ are linearly independent one can conclude that $\{P_k\}$ is total on account of the uniqueness theorem for Fourier series; furthermore, it is known that the linear span of $\bigcup_{m \in Z^n} e^{im\cdot v}$ is fundamental in the above X-spaces.

Thus it remains to determine the domain of α for which (3.7) holds, or, by Theorem 3.19, equivalently for which

$$(5.4) \qquad \|\sum_{k<\rho} (1-\frac{k}{\rho})^\alpha P_k f\| \leq C_\alpha^* \|f\| \qquad (f \in X)$$

is true. Now Bochner has proved that

$$\|\sum_{|m|<\rho'}(1-(\frac{|m|}{\rho'})^2)^\alpha e^{im\cdot v}\|_1 \leq D$$

uniformly in $\rho' > 0$ for $\alpha > (n-1)/2$, but not for $\alpha \leq (n-1)/2$; hence (5.4) holds for all X-spaces provided $\alpha > (n-1)/2$. Furthermore, it is standard that the (radial) partial sums $\sum_{k<\rho} P_k f$, P_k given by (5.3), converge in $L^2(T^n)$ (and only in $L^2(T^n)$, $n \geq 2$, as recently proved by Fefferman [46]). Using interpolation techniques Stein [93] derived from these two results that for all $f \in X$

$$\|\sum_{|m|<\rho'} (1-(\frac{|m|}{\rho'})^2)^\alpha f^\wedge(m)e^{im\cdot v}\|_p \leq D_\alpha \|f\|_p \qquad (\alpha > (n-1)|\tfrac{1}{p}-\tfrac{1}{2}|).$$

Observing that we have set $|m|^2 = k$ (and hence $\rho'^2 = \rho$), it is clear that (5.4) holds for $\alpha > (n-1)|\frac{1}{p} - \frac{1}{2}|$, $1 \leq p \leq \infty$ (in particular (5.4) holds in C-norm for $\alpha > (n-1)/2$). Hence we may apply the results of Sec. 4. In case $\Phi(x) = x^l$, $l \in \mathbb{N}$, Theorem 4.1 is a particular result of Mitjagin [74] and later [90], namely

Corollary 5.5. Let X, $\{P_k\}$ be given as above, let f and all its derivatives up to and including order $2l$ exist and belong to X. Then there exists a polynomial (radial) approximation process $T(\rho)$ with associated multiplier sequences $\{\tau(\rho)\}$ such that

$$\|\sum_{|m|<\rho} \tau_{|m|}(\rho)f^\wedge(m)e^{im\cdot v} - f(v)\| \leq D2^{2l}\rho^{-2l}\|\Delta^l f\|,$$

where $\Delta = \sum_{k=1}^n(\partial^2/\partial v_k^2)$ is the standard Laplacian.

Of course many other choices are possible for Φ (cf.e.g.(3.22)). Analogously the Bernstein-type inequality 4.2 yields, for example,

Corollary 5.6. Let X, $\{P_k\}$ be given as above. Then, for $\beta > 0$, $\gamma = 0$ or $\beta \geq 0$, $\gamma = 1$, $\Phi(x) = x^{\beta/2}\log^\gamma(1+x)$, one has ($c_m \in \mathbb{C}$, $m \in \mathbb{Z}^n$)

$$\|\sum_{|m|<\rho}|m|^\beta \log^\gamma(1+|m|^2)c_m e^{im\cdot v}\| \leq D\rho^\beta \log^\gamma(1+\rho^2)\|\sum_{|m|<\rho}c_m e^{im\cdot v}\| .$$

For even $\beta \in \mathbb{N}$, $\gamma = 0$, statement i) coincides with results of Mitjagin [74] (who also considers non-radial factors, e.g. $\prod_{k=1}^n m_k^{2l_k}$, $l_k \in \mathbb{N}$); in particular the above corollary delivers ($|m|^2 \sim \Delta$)

$$\|\Delta \sum_{|m|<\rho} c_m e^{im\cdot v}\| \leqslant D\rho^2 \|\sum_{|m|<\rho} c_m e^{im\cdot v}\| .$$

To give an example for a comparison theorem, let us apply Theorem 4.6 in connection with (3.43) to obtain

<u>*Corollary 5.7.*</u> *Let X, $\{P_k\}$, and Φ be as in Corollary 5.6. Then the Abel-Cartwright means and Riesz means are equivalent for $\lambda > (n-1)\left|\frac{1}{p} - \frac{1}{2}\right|$, i.e.,*

$$\|W_\Phi(\rho)f - f\| \approx \|R_{\Phi,\lambda}(\rho)f - f\| \qquad (f \in X). \tag{5.5}$$

For $\gamma = 0$, one direction of (5.5) is contained in Löfström [70]; indeed, it is proved there that the Abel-Cartwright means are better than the Riesz means (in fact Löfström proved a far more general result replacing $x^{\beta/2}$ (which corresponds to $|v|^\beta$, $v \in R^n$) by a positive-homogeneous function: $\Phi(cv) = c^\gamma\Phi(v)$, $v \in R^n$ and $c,\gamma > 0$, which is not necessarily radial; however note that $|v|^\beta \log(1+|v|^2)$ does not immediately fall under his scope.) Equally, for homogeneous, radial functions Φ the analogous part of (4.12) is contained in [70], namely that the Abel-Cartwright means are better than the Bessel potentials.

Concerning the saturation problem, Theorem 4.4 iii) with $\Phi(x) = x^\beta$, $\beta > 0$, and Theorems 4.7 iii) and 4.8 iii) with $\Phi(x) = x$ are contained in Nessel-Pawelke [82]. As already mentioned, a saturation theorem for $\mathcal{T}^*$ on $L^p(R^n)$ essentially also gives the saturation theorem for its periodic analog $\mathcal{T}$ on $L^p(T^n)$. Therefore, we list some papers dealing with saturation on $L^p(R^n)$: [18], [30], [55,56], [66,67], [70], [81], etc.; many of these results (in their periodic version) could be obtained by the present approach. To give an example, we have by Theorem 4.8 iii)

<u>*Corollary 5.8.*</u> *Let X, $\{P_k\}$ be given as above, and let $\lambda > (n-1)\left|\frac{1}{p} - \frac{1}{2}\right|$, $1 \leqslant p < \infty$, or $\lambda > (n-1)/2$ in the continuous case. Then for*

$$R_{log,\lambda}(\rho)f(v) \equiv \sum_{|m|<\rho} \left(1 - \frac{log(1+|m|^2)}{log(1+\rho^2)}\right)^\lambda f^\wedge(m)e^{im\cdot v}$$

one has

a) $$\|R_{log,\lambda}(\rho)f - f\| = o(log^{-1}(1+\rho^2)) \qquad (\rho \to \infty)$$

implies f is a constant;

b) $$\| R_{log,\lambda}(\rho)f - f\| = O(\log^{-1}(1+\rho^2)) \qquad (\rho \to \infty)$$

if and only if

$$\| \sum_{|m|<\rho} (1 - \frac{|m|}{\rho})^\lambda \log(1+|m|^2) f^\wedge(m) e^{im\cdot v}\| = O(1) \qquad (\rho \to \infty).$$

To give an idea how (little) smooth the functions in b) are we mention that $\Delta^{2l} \sim |m|^{2l}$, $l \in \mathbb{N}$, Δ being the Laplacian.

Finally, to give an example for a Zamansky-type inequality we state

Corollary 5.9. *Let X and $\{P_k\}$ be given as above, set*

$$R_{\beta,\lambda}(\rho)f(v) \equiv \sum_{|m|<\rho} \left(1 - (\frac{|m|}{\rho})^\beta\right)^\lambda f^\wedge(m) e^{im\cdot v} \qquad (f \in X).$$

Then for all $\lambda > (n-1)|\frac{1}{p} - \frac{1}{2}|$, $\gamma \geq \beta$, and each $f \in X$

$$\| \sum_{|m|<\rho} \left(1 - (\frac{|m|}{\rho})^\beta\right)^\lambda |m|^\gamma f^\wedge(m) e^{im\cdot v}\| \leq D\rho^\gamma \| R_{\beta,\lambda}(\rho)f - f\| \quad .$$

Remarks. a) It is clear that an improvement of the fundamental result (5.4) for $\lambda > (n-1)|1/p - 1/2|$ (which we need as a basic hypothesis) would give further multiplier and approximation results for the n-dimensional trigonometric system via Sec. 3 and 4. Therefore, let us mention two recent results.

i) Fefferman [45] has shown that for $n \geq 2$

$$\| \sum_{|m|<\rho} \left(1 - (\frac{|m|}{\rho})^2\right)^\lambda f^\wedge(m) e^{im\cdot v}\|_p \leq D\| f\|_p \qquad (f \in L^p(T^n))$$

provided $1 < p < 4n/(3n+1)$ and $\lambda > \frac{n}{p} - \frac{n+1}{2}$.

ii) Carleson-Sjölin [38] have improved (5.4) for $n = 2$ in another direction, namely to

$$\| \sum_{|m|<\rho} \left(1 - (\frac{|m|}{\rho})^2\right)^\lambda f^\wedge(m) e^{im\cdot v}\|_p \leq D\| f\|_p \qquad (f \in L^p(T^2))$$

provided $4/3 < p < 4$ and $\lambda > 0$. Finally, according to a written communication of P. Sjölin, Fefferman's result may be further improved to $1 < p \leq 4/3$ by making use of the estimate of the operator R mentioned at the end of the introduction in [38].

b) Our results on $L^p(T^n)$ are also valid on $L^p(R^n)$ with the corresponding approximation processes. For, setting $x = |v|$, the functions $e(x) \in BV_{\alpha+1}$ (and naturally $e(\Phi(x))$, Φ appropriate) are multipliers on $L^p(R^n)$ provided $\alpha > (n-1)|1/p - 1/2|$. Here a measurable function e is a multiplier on $L^p(R^n)$ or a multiplier of type $(L^p(R^n), L^p(R^n)) \equiv (L^p, L^p)$ iff i) in case $p = 1$, $p = \infty$ there exists a bounded Borel measure h on R^n such that e is the Fourier-Stieltjes transform of that measure almost everywhere, i.e., $(u \in R^n)$

$$e(v) = (2\pi)^{-n/2} \int_{R^n} e^{-iv\cdot u}\, dh(u) \qquad \text{a.e.};$$

ii) in case $1 < p < \infty$ the map $f \to \mathfrak{F}^{-1}[ef^\wedge]$ is bounded in L^p for all f of some appropriate dense subset of L^p; here $f^\wedge$ is the standard Fourier transform of f, i.e.,

$$f^\wedge(v) = (2\pi)^{-n/2} \int_{R^n} f(u) e^{-iv\cdot u} du$$

and $\mathfrak{F}^{-1}$ is the inverse operator.

The proof that $e \in BV_{\alpha+1}$ implies $e(|v|) \in (L^p, L^p)$ for α, p as above is (naturally) based upon the observation that

$$(5.6) \quad r_\alpha(v) = \begin{cases} (1-|v|)^\alpha, & 0 \leq |v| \leq 1 \\ 0, & |v| \geq 1 \end{cases} \in (L^p, L^p), \quad \alpha > (n-1)\left|\frac{1}{p} - \frac{1}{2}\right|$$

(see [70], [93]). Using the (fractional) integral calculus developed in Sec. 3.2 and 3.3 it is obvious that

$$e(|v|) = \frac{\pm 1}{\Gamma(\alpha+1)} \int_{|v|}^{\infty} \left(1 - \frac{|v|}{|u|}\right)^\alpha |u|^\alpha \, de^{(\alpha)}(|u|),$$

and hence for arbitrary f of a sufficiently smooth, dense subspace of L^p $(z \in R^n)$

$$\| \mathfrak{F}^{-1}[e(|v|)f^\wedge(v)](z)\|_p = \| (2\pi)^{-n/2} \int_{R^n} e(|v|) f^\wedge(v) e^{iz\cdot v} dv \|_p$$

$$= \left\| \frac{1}{\Gamma(\alpha+1)} \int_0^\infty |u|^\alpha de^{(\alpha)}(|u|) (2\pi)^{-n/2} \int_{|v|\leq|u|} \left(1 - \frac{|v|}{|u|}\right)^\alpha f^\wedge(v) e^{iz\cdot v} dv \right\|_p$$

$$\leq C_\alpha \|e\|_{BV_{\alpha+1}} \|f\|_p \qquad \left(\alpha > (n-1)\left|\frac{1}{p} - \frac{1}{2}\right|\right)$$

by (5.6) and the generalized Minkowski-inequality. Thus we have proved (for $p = 1$ and $\alpha \in \mathbb{N}$ see [103])

Theorem 5.10. If $e \in BV_{\alpha+1}$ and Φ satisfies the conditions of Lemma 3.21, then

$$e(\Phi(|v|)) \in (L^p(\mathbb{R}^n),\ L^p(\mathbb{R}^n)) \qquad (\alpha > (n-1)|\tfrac{1}{p} - \tfrac{1}{2}|).$$

This result gives a partial (since only radial e are admitted) solution of the following problem of Stein [96;p.110] : give sufficient conditions for a multiplier to belong to some (L^p, L^p), $p \neq 2$, without implying also that it belongs to (L^p, L^p), $1 < p < \infty$.

c) The minor observations to the various results of this subsection have made it clear that the crucial point of our general approach with respect to an application to multiple Fourier series is to be seen in the fact that the multiplier sequences have in principle to be radial. How does one remove this difficulty?

One possibility to deal with functions e(v), even in each coordinate, is to introduce a sequence of n-parameter projections $\{P_{\overline{k}}\}$, $\overline{k} = (k_1,\dots,k_n)$ with $k_i \in P$ (if e(v) is arbitrary one would have to consider two-a-way sequences: $k_i \in Z$) and, analogously to Moore [76] for numerical divergent series, to develop a corresponding multiplier theory.

But from this point of view there is no solution of Stein's problem since we have the uniform boundedness of the rectangular partial sums for all p, $1 < p < \infty$, i.e.,

$$\| \int_{|v_k| \leq \rho_k} f^\wedge(v)\, e^{iu\cdot v} dv \|_p \leq A_p \| f \|_p \qquad (\rho_k > 0,\ 1 \leq k \leq n)$$

(f belonging to an appropriate dense subspace of L^p; see [96;p.100]), and the Marcinkiewicz-Mikhlin-Hörmander multiplier theorems seem to be nearly best possible in this situation.

However, if the function e to be discussed is "partially radial", e.g. $e(v) = e((v_1^2 + v_2^2)^{1/2}, v_3,\dots,v_n)$ or other combinations in two or more variables, it should not be too hard, by using (5.6), to derive multiplier criteria analogous to Theorem 5.10 which give further partial solutions of Stein's problem.

5.3 Laguerre and Hermite series

Let $X = L^p(0,\infty)$, $1 \leq p < \infty$, with $\|f\|_p = (\int_0^\infty |f(x)|^p dx)^{1/p}$, and consider the Laguerre polynomials $L_k^{(\alpha)}$ of order $\alpha > -1$ defined by

$$k!\, L_k^{(\alpha)}(x) = e^x\, x^{-\alpha}(d/dx)^k(e^{-x}\, x^{k+\alpha}) \qquad (k \in P).$$

It is known that $\{\varphi_k^{(\alpha)}\}$,

$$\varphi_k^{(\alpha)}(x) = \{\Gamma(\alpha+1)A_k^\alpha\}^{-1/2}\, x^{\alpha/2} e^{-x/2}\, L_k^{(\alpha)}(x),$$

is an orthonormal system on $(0,\infty)$. Thus the projections

$$P_k^{(\alpha)} f(x) = \left[\int_0^\infty f(y)\varphi_k^{(\alpha)}(y)dy\right]\varphi_k^{(\alpha)}(x)$$

are mutually orthogonal. Furthermore, Askey-Wainger [7] have shown for $\alpha > 0$ and Muckenhoupt [78] for $\alpha > -1$ that the partial sums converge in L^p-norm provided $|1/p - 1/2| < \min\{1/4, (1+\alpha)/2\}$, and recently Poiani [83] proved the analog for the (C,1)-means for $1 \leq p < \infty$ if $\alpha > 0$, and for $2/(2+\alpha) < p < -2/\alpha$ if $-1 < \alpha \leq 0$. In case $-1 < \alpha < 0$ this p-domain cannot be extended with the aid of higher Cesàro means. Following an argument due to Muckenhoupt assume $f \in L^p$ has an expansion $\sum P_k^{(\alpha)} f$. Then, by the Riesz representation theorem, $\|\varphi_k^{(\alpha)}\|_{p'}$ has to be finite; on the other hand, by (3.14), $\|\varphi_k^{(\alpha)}\|_p$ is finite if some (C,j)-means are bounded. Since $\varphi_k^{(\alpha)}(x) = O(x^{\alpha/2})$ for $x \to 0+$ [100; p.101] one obtains the preceding domain $2/(2+\alpha) < p < -2/\alpha$.
Summarizing, the theory of Sec. 4 may be applied.

Concerning Bernstein-type inequalities we note that the $\varphi_k^{(\alpha)}$ are eigenfunctions of the differential operator

$$D_x^{(\alpha)} = \frac{d}{dx}\left(x\frac{d}{dx}\right) + \frac{\alpha+1}{2} - \frac{x}{4} - \frac{\alpha^2}{4x}$$

with eigenvalues $-k$, $k \in P$. Hence Theorem 4.2 yields in particular (see [59;II])

Corollary 5.11. *Let $X, (P_k^{(\alpha)}), \alpha$, and p be given as above. Then, for arbitrary $c_k \in C$,*

$$\| D_x^{(\alpha)} \textstyle\sum_{k=0}^n c_k \varphi_k^{(\alpha)}\| \leq C\, n \|\sum_{k=0}^n c_k\varphi_k^{(\alpha)}\| \qquad (n \in N),$$

the constant C being independent of n and $\{c_k\}$.

Analogously we have for example by Theorem 4.4 iii)

Corollary 5.12. *Let* X, $\{P_k^{(\alpha)}\}$, α , *and* p *be given as above and choose* $\Phi(x) = x^\beta$, $\beta > 0$

a) $$\|\sum_{k=0}^{\infty} e^{-\varepsilon k^\beta} P_k^{(\alpha)} f(x) - f(x)\| = o(\varepsilon) \qquad (\varepsilon \to 0+)$$

implies $f(x) = c\, x^{\alpha/2}\, e^{-x/2}$ *for some constant* c.

b) $$\|\sum_{k=0}^{\infty} e^{-\varepsilon k^\beta} P_k^{(\alpha)} f(x) - f(x)\| = O(\varepsilon) \qquad (\varepsilon \to 0+)$$

if and only if

$$\|\sum_{k=0}^{n} \left(1 - \frac{k}{n+1}\right) k^\beta\, P_k^{(\alpha)} f\| = O(1) \qquad (n \to \infty). \tag{5.7}$$

For $\beta = 1, \alpha = 0$ this result is already proved in Butzer [23] by semi-group methods, who also gives a characterization of (5.7) in case $1 < p < \infty$ which we now derive. Observe that (5.7) is equivalent to $|f|_{\psi^\sim} < \infty$, $\psi = \{k\}$, by Theorem 2.6 a) and that one has $|f|_{\psi^\sim} = |f|_\psi$ in the reflexive case $1 < p < \infty$. Hence, to each $f \in X^\psi$ there exists a function $g \in L^p$ with $B^\psi f = g$. Thus, on account of $B^\psi = D_x^{(\alpha)}$, $\psi = \{k\} \in s$, (5.7) is equivalent to the statement $D_x^{(\alpha)} f(x) \in L^p(0,\infty)$.

Of course it is easy to state a comparison theorem (see [33;I]) or a Zamansky-type theorem, but we omit this.

Remark. Let us mention that Poiani [83] established far more than we stated above; indeed she has proved the (C,1)-boundedness of the Laguerre series expansion in weighted L^p-spaces, $1 \leq p \leq \infty$, where the weight is connected with the domains of α and p. Of course the results of Sec. 3 and 4 could have been applied to these weighted spaces. In contrast we note that Muckenhoupt [78] has shown the boundedness of the partial sums for $1 < p < \infty$ with a stronger weight on the right hand side than on the left hand side (see the Remark following Theorem 3.3).

Concerning the Hermite series case choose $X = L^p(-\infty,\infty)$, $1 \leq p < \infty$, with $\|f\|_p = (\int_{-\infty}^{\infty} |f(x)|^p dx)^{1/p}$ finite, and consider the Hermite polynomials defined by

$$H_k(x) = (-1)^k\, e^{x^2} (d/dx)^k\, e^{-x^2} \qquad (k \in P).$$

Setting

$$\varphi_k(x) = (2^k\ k!\ \sqrt{\pi})^{-1/2}\ e^{-x^2/2}\ H_k(x),$$

$\{\varphi_k\}$ is an orthonormal family of functions on $(-\infty,\infty)$. Thus the projections

$$P_k f(x) = \left[\int_{-\infty}^{\infty} f(y)\varphi_k(y)dy\right] \varphi_k(x)$$

are mutually orthogonal. Again as in the Laguerre series case, Askey-Wainger [7] proved that the partial sums converge in mean provided $4/3 < p < 4$, whereas Poiani [83] has shown that the (C,1)-means converge in norm provided $1 < p < \infty$ (or for appropriate weighted L^p-spaces, $1 \leq p < \infty$). Now observing that the φ_k are eigenfunctions of the differential operator $(d^2/dx^2) + (1-x^2)$ with eigenvalues $-2k$, $k \in P$, Theorem 4.2 yields

Corollary 5.13. *For arbitrary* $c_k \in C$ *one has*

$$\| \{(\tfrac{d}{dx})^2 + (1-x^2)\}\sum_{k=0}^{n} c_k\varphi_k(x)\|_p \leq C\, n \|\sum_{k=0}^{n} c_k\varphi_k(x)\|_p \qquad (1 < p < \infty),$$

the constant C *being independent of* n *and* $\{c_k\} \in s$.

This inequality is contained in a paper of Freud [47] , whose methods cover $p = 1$ and $p = \infty$ as well. If the Hermite expansion is "regular" in Stein's [94] terminology for all $1 \leq p \leq \infty$, i.e., if the expansion is (C,j)-bounded for some $j \in P$ and all $1 \leq p \leq \infty$ and thus Cor. 5.13 also holds for $p = 1$ and $p = \infty$, remains open since the (negative) argument for the Laguerre series case seems not to apply immediately.

Theorem 4.4 iii) for $\Phi(x) = x$ is again contained in [23] with the characterization of the relative completion by $\{(d/dx)^2+(1-x^2)\}f(x) \in L^p$ in the reflexive case $1 < p < \infty$. (Indeed, Butzer [23] also derives a Jackson-type inequality in the Hermite as well as in the Laguerre series case for $p = 1$; all his results are not based on our fundamental property (3.7)).

Let us conclude this subsection with a comparison example.

Corollary 5.14. *Let* $f \in L^p(-\infty,\infty)$, $1 < p < \infty$, *and* $\{P_k\}$ *as above. Then*

$$\| \sum_{k<\rho}(1 -(\tfrac{k}{\rho})^{\beta})^{\lambda} P_k f - f\|_p \leq C \| \sum_{k<\rho} (1 -(\tfrac{k}{\rho})^{\gamma})^{\lambda} P_k f - f\|_p ,$$

where $\lambda \geq 1$, $0 < \gamma < \beta$, *and* C *being independent of* ρ *and* f.

This example is implicitly contained in [33;II]; obviously we could also have taken another choice of Φ in Theorem 4.7 v .

5.4 Jacobi series

Let X be the Banach space of all measurable functions, continuous on $[-1,1]$, i.e.,

$$C[-1,1] = \{f;\ \|f\|_C = \max_{-1\leq x\leq 1} |f(x)| < \infty\},$$

or p-th power integrable on $(-1,1)$ with respect to the weight $(1-x)^\alpha(1+x)^\beta$, $\alpha,\beta > -1$, i.e.,

$$L^p_{(\alpha,\beta)}(-1,1) = \{f; \|f\|_p = (\int_{-1}^{1} |f(x)|^p(1-x)^\alpha(1+x)^\beta dx)^{1/p} < \infty,\ 1\leq p < \infty\}.$$

Let $B_k^{(\alpha,\beta)}(x)$ be the Jacobi polynomial of degree k, order (α,β) defined by $(\alpha,\beta > -1)$

$$(1-x)^\alpha(1+x)^\beta\ B_k^{(\alpha,\beta)}(x) = \frac{(-1)^k}{2^k k!}(\frac{d}{dx})^k\{(1-x)^{k+\alpha}(1+x)^{k+\beta}\}.$$

$B_k^{(\alpha,\beta)}(x)$ are orthogonal on $(-1,1)$ with respect to $(1-x)^\alpha(1+x)^\beta$ and

$$\int_{-1}^{1} \{B_k^{(\alpha,\beta)}(x)\}^2(1-x)^\alpha(1+x)^\beta dx$$

$$= \frac{2^{\alpha+\beta+1}\ \Gamma(k+\alpha+1)\ \Gamma(k+\beta+1)}{(2k+\alpha+\beta+1)\ \Gamma(k+\alpha+\beta+1)\ \Gamma(n+1)} = \{h_k^{(\alpha,\beta)}\}^{-1} .$$

Thus the projections

$$P_k^{(\alpha,\beta)}f(x) = [\int_{-1}^{1} f(y)B_k^{(\alpha,\beta)}(y)(1-y)^\alpha(1+y)^\beta dy]\ h_k^{(\alpha,\beta)}B_k^{(\alpha,\beta)}(x)$$

are mutually orthogonal. Furthermore, the expansion $\sum P_k f$ is unique and the linear span $\bigcup_{k\in P} P_k(X)$ is dense in the above X-spaces. Pollard [84] has shown for $\beta,\alpha \geq -1/2$ and Muckenhoupt [77] for $\alpha,\beta > -1$ that the partial sums $(C,0)_n f$ converge to f provided

(5.8) $\quad (\alpha+1)|\frac{1}{p} - \frac{1}{2}| < \min(\frac{1}{4}, \frac{1+\alpha}{2}) \quad$ and $\quad (\beta+1)|\frac{1}{p} - \frac{1}{2}| < \min(\frac{1}{4}, \frac{1+\beta}{2})$, and there is no norm convergence if p lies outside one of these ranges. In order to deduce the uniform boundedness of the operators $(C,\alpha)_n$ for appropriate $\alpha > 0$, the convolution structure for Jacobi polynomials as developed by Askey-Wainger [9] is basic. First it is shown that there exists a function $g(x;y) \in X$ such that

$$g(x;y) \sim \sum_{k=0}^{\infty} P_k^{(\alpha,\beta)} g(x)\; B_k^{(\alpha,\beta)}(y)/B_k^{(\alpha,\beta)}(1) \qquad (\alpha \geq \beta) \tag{5.9}$$

and, for some constant $D \geq 1$,

$$\| g(\cdot\, ;y)\| \leq D\, \| g\| \qquad (\alpha \geq \beta \text{ and } \alpha + \beta \geq -1).$$

Then defining

$$f \# g(x) = \int_{-1}^{1} g(x;y)f(y)(1-y)^{\alpha}(1+y)^{\beta}dy$$

for $f \in X$ and $g \in L^1_{(\alpha,\beta)}$ one obtains $\| f \# g\|_X \leq D\| f\|_X \| g\|_1$ and

$$P_k^{(\alpha,\beta)}(f \# g)(x) = \int_{-1}^{1} g(y)B_k^{(\alpha,\beta)}(y)(1-y)^{\alpha}(1+y)^{\beta}dy\; P_k^{(\alpha,\beta)}f(x).$$

Choosing g in particular as

$$S_n^{\gamma}(x) = \sum_{k=0}^{n} (A_{n-k}^{\gamma}/A_n^{\gamma})h_k^{(\alpha,\beta)}B_k^{(\alpha,\beta)}(1)B_k^{(\alpha,\beta)}(x),$$

it is known [100;p.258] that $\| S_n^{\gamma}\|_1$ is uniformly bounded in n provided $\gamma > \alpha + 1/2$, $\alpha \geq \beta$. Since

$$S_n^{\gamma} \# f(x) = \sum_{k=0}^{n} (A_{n-k}^{\gamma}/A_n^{\gamma})P_k^{(\alpha,\beta)}f(x) \equiv (C,\gamma)_n f(x)$$

one obtains that the (C,γ)-means, $\gamma > \alpha + 1/2$, are uniformly bounded on all X-spaces. Interpolating (between $p = 1$ and $p = 2$ or $p = 2$ and $p = \infty$) finally gives for all $f \in L^p_{(\alpha,\beta)}$, $1 \leq p \leq \infty$ ($L^{\infty}_{(\alpha,\beta)}$ is the set of all essentially bounded functions on $(-1,1)$)

$$\| (C,\gamma)_n f\|_p \leq C_{\gamma}\| f\|_p \qquad (\gamma > (\alpha+1/2)\,|1 - \tfrac{2}{p}|,\ 1 \leq p \leq \infty), \tag{5.10}$$

C_{γ} being independent of n and f, provided $\alpha \geq \beta > -1$ with $\alpha + \beta \geq -1$.

For all these facts see Askey-Wainger [9], Askey-Hirschman [5], Gasper [48,49].

Remarks. a) By using the well-known relation [100; p. 59]

$$B_k^{(\alpha,\beta)}(x) = (-1)^k\, B_k^{(\beta,\alpha)}(-x)$$

(5.10) also holds for $\beta \geq \alpha > -1$, $\alpha + \beta \geq -1$ (see Askey-Wainger [9],[11;p. 8]).

b) (5.10) could be improved when Pollard's result (5.8) is used for interpolation.

Now we may apply our previous general results. In this context let us mention that Theorem 3.7 even in the present concrete situation is a simplification of a multiplier criterium for Jacobi series as given by Bavinck [10] or of a corresponding one for ultraspherical series ($\alpha = \beta$) due to Askey-Wainger [6]; it seems that there was and still is a need for sharper multiplier criteria which can be checked easily.

Turning to the approximation results of Sec. 4 we first observe that the Jacobi polynomials $B_k^{(\alpha,\beta)}$ are eigenfunctions of the differential operator

$$D_x^{(\alpha,\beta)} \equiv (1-x^2)(\tfrac{d}{dx})^2 + \{(\beta - \alpha) - (\alpha + \beta + 2)x\}\tfrac{d}{dx}$$

with eigenvalues $-k(k+\alpha+\beta+1)$ (see [100; p. 60]). Choosing now $\Phi(x) = \{x(x+\alpha+\beta+1)\}^{\delta}$, $\delta > 0$, Theorem 4.2 yields

Corollary 5.15. *Let* X, $B_k^{(\alpha,\beta)}$, $\alpha \geq \beta$ *with* $\alpha + \beta > -1$ *be as above. Then, for arbitrary* $c_k \in \mathbb{C}$ *and* $\delta > 0$,

$$\| \sum_{k=0}^{n} \{k(k+\alpha+\beta+1)\}^{\delta} c_k B_k^{(\alpha,\beta)} \| \leq D\{n(n+\alpha+\beta+1)\}^{\delta} \| \sum_{k=0}^{n} c_k B_k^{(\alpha,\beta)} \| ,$$

D being independent of n and $\{c_k\}$.

For $\delta \in \mathbb{N}$ this result is due to Stein [94], who also needs (5.10) to be valid for some $\gamma \geq 0$ (cf. Theorem 4.2); for arbitrary $\delta > 0$ it is due to Bavinck [11;p.55] (but of course a number of other Φ's are admitted here). Furthermore, analogously to [35;p.138], Bavinck [11;p.60] derives a Bernstein inequality of type (4.24) ($l = 2$) for the de La Vallée-Poussin means via an appropriate kernel with Fourier-Jacobi coefficients

$$\frac{\Gamma(n+1)\ \Gamma(n+\alpha+\beta+2)}{\Gamma(n-k-1)\ \Gamma(n+k+\alpha+\beta+2)},$$

the latter being different from our coefficients $\omega_k(n)$ (see (4.16));

however, an (unpleasant) calculation should show that both results (i.e. Lemma 5.7.2 in [11] and Theorem 4.12 here) are equivalent.

Concerning saturation results we mention that Theorem 4.4 iii) coincides with [11;p.51,p.52] if $\Phi(x) = \{x(x+\alpha+\beta+1)\}^{\gamma}$, $0 < \gamma \leq 1$, and with [11;p.53,p.54] if $\Phi(x) = x^{\gamma}$, $0 < \gamma \leq 1$, in particular for $\gamma = 1$, $\alpha = \beta = 0$, with a result of [25]; further, the saturation classes for these two Φ are identical as proved in [11;p.36] which follows here by Lemma 4.5 when setting $c = \alpha + \beta + 1$. Finally, Theorems 4.9 and 4.11 coincide with [11;p.57] and [11;p.59], respectively. (Obviously, the non-optimal approximation results of [11;p.49 - 61] are also contained here by an application of Theorem 1.2.) For the ultraspherical case, i.e., $\alpha = \beta \geq -1/2$, see also the results in [33;I,II], [59;II].

We conclude this section with a particular instance of Theorem 4.7 iv) and (4.12).

Corollary 5.16. *Let X, $P_k^{(\alpha,\beta)}$, α, and β be as above. Then, for $\gamma \geq \delta > 0$ and $\lambda > (2\alpha+1)|1/p - 1/2|$, $1 \leq p < \infty$, or $\lambda > (\alpha+1/2)/2$ in the continuous case,*

$$\| \textstyle\sum_{k<\rho}(1-(k/\rho)^{\delta})^{\lambda}\, k^{\gamma} P_k^{(\alpha,\beta)} f \|$$

$$\leq D\rho^{\gamma}\| \textstyle\sum_{k<\rho}(1-(k/\rho)^{\delta})^{\lambda}\, P_k^{(\alpha,\beta)} f - f\| \qquad (f \in X)$$

or in the notation following Corollary 5.3

$$\| D_x^{(\alpha,\beta)} R_{\delta,\lambda} f\|_p \leq D\,\rho^{2}\| R_{\delta,\lambda} f - f\|_p \qquad (0 < \delta \leq 1; f \in X).$$

Corollary 5.17. *Let X, $P_k^{(\alpha,\beta)}$, α, and β be as above; choose e.g. $\Phi(x) = \{x(x+\alpha+\beta+1)\}^{\delta}$, $\delta > 0$. Then the Abel-Cartwright means (4.3) and the Bessel potentials (4.10) are equivalent in the sense of Def. 1.4.*

5.5 Surface spherical harmonics

Let Ω_n be the surface of the unit sphere in R^n with elements u', v', content $|\Omega_n| = 2\pi^{n/2}/\Gamma(n/2)$ and surface element ds. Now X denotes one of the spaces $L^p(\Omega_n)$, $1 \leq p < \infty$, or $C(\Omega_n)$ with the usual norms

$$\|f\|_p = \{|\Omega_n|^{-1} \int_{\Omega_n} |f(u')|^p ds(u')\}^{1/p} \quad (1 \leq p < \infty), \quad \|f\|_C = \max_{u' \in \Omega_n} |f(u')|,$$

respectively. If $Y_k(u)$ is a homogeneous polynomial of degree k in n dimensions which satisfies

$$\Delta Y_k(u) = 0, \qquad \Delta = \sum_{k=1}^{n} (\partial/\partial u_k)^2 \qquad (u \in R^n),$$

then the restriction of Y_k to Ω_n, denoted again by Y_k, is called a spherical harmonic of order k. Now it is known that every spherical harmonic of degree k and dimension n satisfies

$$\tilde{\Delta} Y_k(u') = -k(k+n-2) Y_k(u'), \; \tilde{\Delta} f(u) = |u|^2 \Delta f(u/|u|);$$

for each k there exist

$$H(k,n) = (2k+n-2) \frac{(k+n-3)!}{k! \, (n-2)!}$$

linearly independent surface spherical harmonics of degree k; the set $\{Y_k^i(u'); 1 \leq i \leq H(k,n), k \in P\}$ is fundamental and may be assumed to be orthonormal in X. Thus the projections

$$P_k f(u') = \sum_{i=1}^{H(k,n)} \Big(\int_{\Omega_n} f(v') Y_k^i(v') ds(v')\Big) Y_k^i(u') \equiv Y_k(f;u')$$

are mutually orthogonal. Furthermore, the expansion $\sum P_k f$ is unique and the linear span $\bigcup_{k \in P} P_k(X)$ is dense in the above X-spaces. The (C,α)-means are bounded in all X-spaces provided $\alpha > (n-2)/2$. (Since there is mean convergence in case $p = 2$, one may interpolate between $p = 1$ and $p = 2$ or $p = 2$ and $p = \infty$ as performed in [5]; but we omit this.) For all these properties one may consult the monograph of Müller [80], the paper of Berens-Butzer-Pawelke [16], and the literature cited there.

Now we may state the results of Sec. 3 and 4 for this particular type of expansion. Theorem 4.2 immediately yields

Corollary 5.18. *Let X and Y_k be as above. Then*

$$\| \tilde{\Delta}^l \sum_{k=0}^{n} Y_k \| \leq D \, n^{2l} \| \sum_{k=0}^{n} Y_k \| \qquad (l \in N),$$

the constant D being independent of n and the particular sequence $\{Y_k\}$.

This result is e.g. contained in Butzer-Johnen [28]; obviously, we could also reformulate Theorem 4.4 ii), Theorem 4.7 ii), Theorem 4.8 ii), and Theorem 4.12, parts of these results coinciding with [59;II] but we omit this.

Concerning saturation theorems we remark that Theorem 4.4 iii) for $\Phi(x) = \{x(x+(n-2)/2)\}^{\gamma}$, $\gamma > 0$ or $\Phi(x) = x^{2\gamma}$, $\gamma > 0$ and Theorem 4.9 imply or coincide with results in Berens-Butzer-Pawelke [16;p.229, p.251; p.246,p.252;p.248], respectively. Observe that

$$|\Omega_n|^{-1} \int_{\Omega_n} w_t(u'\cdot v')f(v')ds(v')$$

with ($\lambda = n - 5/2$)

$$w_t(\cos\Theta) = \sum_{k=0}^{\infty} \exp\{-k(k+n-2)t\} \frac{k+n-2}{n-2} B_k^{(\lambda,\lambda)}(\cos\Theta)$$

is called the singular integral of Gauss-Weierstrass on X; it solves the heat equation $(\partial/\partial t)U(u',t) = \tilde{\Delta}\, U(u',t)$ on the sphere with initial value $\lim_{t\to 0+} U(u',t) = f(u')$ for $t > 0$, $u' \in \Omega_n$, and $f \in X$.

The characterization of Favard classes via a representation as a fractional integral in [16;p.259] essentially reduces to an application of Lemma 4.3 with $\Phi(x) = \{x(x+(n-2)/2)\}^{\gamma}$, $\gamma > 0$; the identification of the Favard classes in [16;p.256-7], determined by $\Phi(x) = x^{2\gamma}$, $\gamma > 0$, and $\{x(x+(n-2)/2)\}^{\gamma}$, consists in an application of Lemma 4.5.

Analogously, we could easily state Zamansky-type and comparison theorems.

A conjecture concerning multiplier and summability theory

Considering the approximation theories for summation methods of expansions into Jacobi polynomials and spherical harmonics, as presented in [11] and [16], respectively, the essential deficiency of these theories is to be seen in the lack of a "good" multiplier theory. This lack is partially removed by our general theory in Sec.3. Starting from the hypothesis that the Cesàro means of order α are bounded we developed straightforwardly an abstract multiplier theory which, on the one hand, may certainly be further refined which, on the other hand, is however optimal in the sense that the basic assumption (3.7) may be regained from the multiplier theory. Indeed, recalling the remark following Theorem 3.19, by the uniform boundedness of the (C,α)-means (3.7), one obtains Theorem 3.3; this in turn implies Theorem 3.18; an application of Theorem 3.18 delivers the uniform boundedness of the Riesz means of order α, the latter implying again the starting point of our chain (3.7) by Theorem 3.19. Thus we have a type of equivalence between (C,α)-boundedness of the series and the corresponding multiplier theory for Fourier expansions in Banach spaces.

Considering now applications it is evident that every improvement concerning (C,α)-summability for a particular instance of expansion implies a better multiplier theory in this instance by Sec.3. The converse seems to be more difficult. Indeed, in the concrete situation of the trigonometric or ultraspherical system multiplier criteria are known, e.g. those of the Marcinkiewicz-Mikhlin-Hörmander type (see Muckenhoupt-Stein [79] and [8]), which do not seem to fit into the theory of (C,α)-summability (or into the corresponding multiplier theory).

Hence there remains the interesting question as to how far one can interpret multiplier criteria in terms of summability results or, equivalently, as to how far one can establish a correspondence between summability theory (not necessarily (C,α)-theory) on the one hand, and multiplier theory on the other for concrete orthogonal systems.

REFERENCES

[1] ALEXITS, G., Sur l'ordre de grandeur de l'approximation d'une fonction périodique par les sommes de Fejér, Acta Math. Sci. Hungar. 3 (1952), 29 - 42.

[2] ANDERSEN, A.F., Comparison theorems in the theory of Cesàro summability, Proc. London Math. Soc., 2.Ser. 27 (1928), 39 - 71.

[3] ANDERSEN, A.F., On the extension within the theory of Cesàro summability of a classical convergence theorem of Dedekind, Proc. London Math. Soc.,3.Ser., 8 (1958), 1 - 52.

[4] ASKEY, R., A transplantation theorem for Jacobi series, Illinois J. Math. 13 (1969), 583 - 590.

[5] ASKEY, R. - I.I. HIRSCHMAN, Jr., Mean summability for ultraspherical polynomials, Math.Scand. 12 (1963), 167 - 177.

[6] ASKEY, R. - S. WAINGER, On the behaviour of special classes of ultraspherical expansions I, J.Analyse Math. 15 (1965), 193 - 220.

[7] ASKEY, R. - S. WAINGER, Mean convergence of expansions in Laguerre and Hermite series, Amer.J.Math. 87 (1965), 695 - 708.

[8] ASKEY, R. - S. WAINGER, A transplantation theorem between ultraspherical series, Illinois J.Math. 10 (1966), 322 - 344.

[9] ASKEY, R. - S. WAINGER, A convolution structure for Jacobi series, Amer.J.Math. 91 (1969), 463 - 485.

[10] BAVINCK, H., A special class of Jacobi series and some applications, J.Math.Anal.Appl. 37 (1972), 767 - 797.

[11] BAVINCK, H., Jacobi series and approximation, Thesis, Mathematisch Centrum - Amsterdam 1972.

[12] BENEDEK, A. - R. PANZONE, Note on mean convergence of eigenfunction expansions, Rev.Un.Mat.Argentina 25 (1970), 167 - 184.

[13] BENEDEK, A. - R. PANZONE, On mean convergence of Fourier-Bessel series of negative order, Stud. in Applied Math. 50 (1971), 281 - 292.

[14] BENEDEK, A. - R. PANZONE, Mean convergence of series of Bessel functions, Rev.Un.Mat.Argentina 26 (1972), 42 - 61.

[15] BERENS, H., Interpolationsmethoden zur Behandlung von Approximationsprozessen auf Banachräumen, Lecture Notes in Math. Vol. 64, Springer, Berlin 1968.

[16] BERENS, H. - P.L. BUTZER - S. PAWELKE, Limitierungsverfahren von Reihen mehrdimensionaler Kugelfunktionen und deren Saturationsverhalten, Publ.Res.Inst.Math.Sci.Ser. A 4 (1968), 201 - 268.

[17] BIEBERACH, L., Lehrbuch der Funktionentheorie I, Teubner, Berlin 1921.

[18] BOMAN, J., Saturation problems and distribution theory, in: Lecture Notes in Math.Vol.187, pp.249-266, Springer, Berlin 1971.

[19] BOMAN, J. - H.S. SHAPIRO, Comparison theorems for a generalized modulus of continuity, Ark.Mat.9 (1971), 91 - 116.

[20] BORWEIN, D., A summability factor theorem, J. London Math.Soc. 25 (1950), 302 - 315.

[21] BORWEIN, D., Integration by parts of Cesàro summable integrals, J. London Math.Soc. 29 (1954), 276 - 292.

[22] BOSANQUET, L.S., Note on convergence and summability factors, I J. London Math.Soc. 20 (1945), 39 - 48; II+III Proc. London Math.Soc., 2.Ser., 50 (1949), 295 - 304, 482 - 496.

[23] BUTZER, P.L., Halbgruppen von linearen Operatoren und das Darstellungs- und Umkehrproblem für Laplace-Transformationen, Math.Ann. 134 (1957), 154 - 166.

[24] BUTZER, P.L., Sur le rôle de la transformation de Fourier dans quelques problèmes d'approximation, C.R.Acad.Sci. Paris 249 (1959), 2467 - 2469.

[25] BUTZER, P.L., Integral transform methods in the theory of approximation, in: On Approximation Theory (P.L. Butzer - J. Korevaar, Eds.), pp. 12 - 23, ISNM 5, Birkhäuser, Basel 1964.

[26] BUTZER, P.L. - H. BERENS, Semi-Groups of Operators and Approximation, Springer, Berlin 1967.

[27] BUTZER, P.L. - E. GÖRLICH, Saturationsklassen und asymptotische Eigenschaften trigonometrischer singulärer Integrale, in: Festschrift zur Gedächtnisfeier Karl Weierstraß 1815 - 1965 (H.Behnke - K.Kopfermann, Eds.), pp.339 - 392, Wiss.Abh. Arbeitsgemeinschaft für Forschung des Landes Nordrhein-Westfalen 33, Westdeutsch.Verl., Köln-Opladen 1966.

[28] BUTZER, P.L. - H. JOHNEN, Lipschitz spaces on compact manifolds, J. Functional Analysis 7 (1971), 242 - 266.

[29] BUTZER, P.L. - W. KOLBE - R.J. NESSEL, Approximation by functions harmonic in a strip, Arch. Rational Mech.Anal. 44 (1972), 329 - 336.

[30] BUTZER, P.L. - R.J. NESSEL, Contributions to the theory of saturation for singular integrals in several variables I. General theory, Nederl.Akad.Wetensch.Indag.Math. 28 (1966),515-531.

[31] BUTZER, P.L. - R.J. NESSEL, Fourier Analysis and Approximation, Vol.I, Birkhäuser, Basel und Academic Press, New York 1971.

[32] BUTZER, P.L. - R.J. NESSEL - W. TREBELS, On the comparison of approximation processes in Hilbert spaces, in: Linear Operators and Approximation (P.L. Butzer - J.-P. Kahane - B. Sz.-Nagy, Eds.), pp.234 - 253, ISNM 20, Birkhäuser, Basel 1972.

[33] BUTZER, P.L. - R.J. NESSEL - W. TREBELS, On summation processes of Fourier expansions in Banach spaces I. Comparison theorems, Tôhoku Math.J. 24 (1972), 127 - 140; II Saturation theorems, ibid. 24 (1972) (in print).

[34] BUTZER, P.L. - S. PAWELKE, Ableitungen von trigonometrischen Approximationsprozessen, Acta Sci.Math.(Szeged) 28 (1967), 173 - 183.

[35] BUTZER, P.L. - K. SCHERER, Approximationsprozesse und Interpolationsmethoden, (Hochschulskripten 826/826a) Bibliograph. Inst. 1968.

[36] BUTZER, P.L. - K. SCHERER, Approximation theorems for sequences of commutative operators in Banach spaces, in: Proc. Internat. Conf. on Constructive Function Theory (B.Penkov - D.Vačov, Eds.), pp.137 - 146, Publishing House Bulgarian Acad. Sci., Sofia 1972.

[37] BUTZER, P.L. - K. SCHERER, Jackson and Bernstein-type inequalities for families of commutative operators in Banach spaces, J. Approximation Theory 5 (1972), 308 - 342.

[38] CARLESON, L. - P. SJÖLIN, Oscillatory integrals and a multiplier problem for the disc, Studia Math. 44 (1972), 287 - 299.

[39] CHANDRASEKHARAN, K. - S. MINAKSHISUNDARAM, Typical Means, (Tata Inst. of Fundamental Res., Bombay; Monographs on Math. and Phys. 1) Oxford Univ.Press, Oxford 1952.

[40] CHAPMAN, S., On non-integral orders of summability of series and integrals, Proc. London Math.Soc., 2.Ser., 9 (1911),369-409.

[41] COSSAR, J., A theorem on Cesàro summability, J. London Math.Soc. 16 (1941), 56 - 68.

[42] DAVIS, P.J., Interpolation and Approximation, Blaisdell Publ., New York 1963.

[43] FAVARD, J., Sur la saturation des procédés de sommation, J.Math. Pures Appl. 36 (1957), 359 - 372.

[44] FAVARD, J., On the comparison of the processes of summation, SIAM J. Numer. Anal. Ser.B 1 (1964), 38 - 52.

[45] FEFFERMAN, C., Inequalities for strongly singular convolution operators, Acta Math. 124 (1970), 9 - 36.

[46] FEFFERMAN, C., The multiplier problem for the ball, Ann. of Math. 94 (1971), 330 - 336.

[47] FREUD, G., On an inequality of Markov type, Soviet Math. Dokl. 12 (1971), 570 - 573.

[48] GASPER, G., Positivity and the convolution structure for Jacobi series, Ann. of Math. 93 (1971), 112 - 118.

[49] GASPER, G., Banach algebras for Jacobi series and positivity of a kernel, Ann. of Math. 95 (1972), 261 - 280.

[50] GERGEN, J.J., Summability of double Fourier series, Duke Math.J. 3 (1937), 133 - 148.

[51] GILBERT, J.E., Maximal theorems for some orthogonal series.
I Trans. Amer. Math. Soc. 145 (1969), 495 - 515;
II J. Math. Anal. Appl. 31 (1970), 349 - 368.

[52] GOES, G., Multiplikatoren für starke Konvergenz von Fourierreihen I, II, Studia Math. 17 (1958), 299 - 311.

[53] GOES, G., Charakterisierung von Fourierkoeffizienten mit einem Summierbarkeitsfaktorentheorem und Multiplikatoren, Studia Math. 19 (1960), 133 - 148.

[54] GOES, G. - S. GOES, Sequences of bounded variation and sequences of Fourier coefficients I, Math.Z. 118 (1970), 93 - 102.

[55] GÖRLICH, E., Distributional methods in saturation theory, J. Approximation Theory 1 (1968), 111 - 136.

[56] GÖRLICH, E., Saturation theorems and distributional methods, in: Abstract Spaces and Approximation (P.L. Butzer - B. Sz.-Nagy, Eds.), pp.218 - 232, ISNM 10, Birkhäuser, Basel 1969.

[57] GÖRLICH, E., Logarithmische und exponentielle Ungleichungen vom Bernstein-Typ und verallgemeinerte Ableitungen, Habilitationsschrift, Aachen 1971.

[58] GÖRLICH, E., Logarithmic and exponential variants of Bernstein's inequality and generalized derivatives, in: Linear Operators and Approximation (P.L. Butzer - J.-P. Kahane - B. Sz.-Nagy, Eds.), pp. 325 - 337, ISNM 20, Birkhäuser, Basel 1972.

[59] GÖRLICH, E. - R.J. NESSEL - W. TREBELS, Bernstein-type inequalities for families of multiplier operators in Banach spaces with Cesàro decompositions I General theory, II Applications (to appear in Acta Sci. Math (Szeged)).

[60] HARDY, G.H., The second theorem of consistency for summable series, Proc. London Math. Soc., 2.Ser., 15 (1916), 72 - 88.

[61] HARDY, G.H., Divergent Series, Oxford at the Clarendon Press 1949.

[62] HARŠILADSE, F.I., Saturation classes for some summation processes (Russ.), Dokl. Akad. Nauk SSSR 122 (1958), 352 - 355.

[63] HILLE, E., Remarks on ergodic theorems, Trans. Amer. Math. Soc. 57 (1945), 246 - 269.

[64] HIRST, K.A., On the second theorem of consistency in the theory of summation by typical means, Proc. London Math. Soc., 2.Ser., 33 (1932), 353 - 366.

[65] INGHAM, A.E., The equivalence theorem for Cesàro and Riesz summability, Publ. Ramanujan Inst. 1 (1969), 107 - 113.

[66] KOZIMA, M. - G. SUNOUCHI, On the approximation and saturation by general singular integrals, Tôhoku Math. J. 20 (1968), 146 - 169.

[67] KOZIMA, M. - G. SUNOUCHI, Characterization of saturation classes in R^n (in print).

[68] KUTTNER, B., Note on the "Second Theorem of Consistency" for Riesz summability, I J. London Math. Soc. 26 (1951), 104 - 111; II ibid. 27 (1952), 207 - 217.

[69] LI SJUN-CZIN, Cesàro summability in Banach space, Chinese Math.-Acta 1 (1960), 40 - 52.

[70] LÖFSTRÖM, J., Some theorems on interpolation spaces with applications to approximation in L_p, Math. Ann. 172 (1967), 176-196.

[71] MADDOX, I.J., Convergence and summability factors for Riesz means, Proc. London Math. Soc. 12 (1962), 345 - 366.

[72] MARTI, J.T., Introduction to the Theory of Bases, Springer, Berlin 1969.

[73] MILMAN, V.D., Geometric theory of Banach spaces I, The theory of bases and minimal systems, Russian Math. Surveys 25 (1970), 111 - 170.

[74] MITJAGIN, B.S., Approximation von Funktionen im Raume L^p und C auf dem Torus (Russ.), Mat. Sb. 58(4) (1962), 397 - 414.

[75] MOORE, C.N., On the use of Cesàro means in determining criteria for Fourier constants, Bull. Amer.Math.Soc.39 (1933), 907-913.

[76] MOORE, C.N., Summable Series and Convergence Factors, Dover Publ., New York 1966.

[77] MUCKENHOUPT, B., Mean convergence of Jacobi series, Proc. Amer. Math. Soc. 23 (1969), 306 - 310.

[78] MUCKENHOUPT, B., Mean convergence of Hermite and Laguerre series II, Trans. Amer. Math. Soc. 147 (1970), 433 - 460.

[79] MUCKENHOUPT, B. - E.M. STEIN, Classical expansions and their relation to conjugate harmonic functions, Trans.Amer.Math.Soc. 118 (1965), 17 - 92.

[80] MÜLLER, C., Spherical Harmonics, Lecture Notes in Math. Vol. 17, Springer, Berlin 1966.

[81] NESSEL, R.J., Contributions to the theory of saturation for singular integrals in several variables II Applications, III Radial kernels, Nederl.Akad.Wetensch.Indag.Math. 29 (1967), 52 - 73.

[82] NESSEL, R.J. - A. PAWELKE, Über Favardklassen von Summationsprozessen mehrdimensionaler Fourierreihen, Compositio Math. 19 (1968), 196 - 212.

[83] POIANI, E.L., Mean Cesàro summability of Laguerre and Hermite series and asymptotic estimates of Laguerre and Hermite polynomials, Thesis, Rutgers Univ., New Brunswick, N.J., 1971.

[84] POLLARD, H., Mean convergence for orthogonal series, III Duke Math. J. 16 (1949), 189 - 191.

[85] RIESZ, M., L'integrale de Riemann-Liouville et le problème de Cauchy, Acta Math. 81 (1948), 1 - 223.

[86] RUSSEL, D.C., On generalized Cesàro means of integral order, Tôhoku Math.J. 17 (1965), 410 - 442.

[87] RYSHIK, I.M. - I.S. GRADSTEIN, Summen-, Produkt- und Integraltafeln, Deutsch.Verl.Wiss., Berlin 1963.

[88] RYCHLÉK, K., Über de La Vallée - Poussins Summationsmethode (Böhmisch), Casopis pro Pěstovani Mathematiky a Fysiky 46 (1917), 313 - 331.

[89] SHAPIRO, H.S., Some Tauberian theorems with applications to approximation theory, Bull. Amer. Math. Soc. 74 (1968), 500 - 504.

[90] SHAPIRO, H.S., Approximation by trigonometric polynomials to periodic functions of several variables, in: Abstract Spaces and Approximation (P.L. Butzer - B. Sz.-Nagy, Eds.), pp. 203 - 217, ISNM 10, Birkhäuser, Basel 1969.

[91] SINGER, I., Bases in Banach Spaces I, Springer, Berlin 1970.

[92] STARK, E.L., Über trigonometrische singuläre Faltungsintegrale mit Kernen endlicher Oszillation, Dissertation, Aachen 1970.

[93] STEIN, E.M., Interpolation of linear operators, Trans.Amer.Math. Soc. 83 (1956), 482 - 492.

[94] STEIN, E.M., Interpolation in polynomial classes and Markoff's inequality, Duke Math.J. 24 (1957), 467 - 476.

[95] STEIN, E.M., The characterization of functions arising as potentials, Bull. Amer. Math. Soc. 67 (1961), 102 - 104.

[96] STEIN, E.M., Singular Integrals and Differentiability Properties of Functions, Princeton Univ. Press, Princeton N.J. 1970.

[97] STEIN, E.M. - G. WEISS, Introduction to Fourier Analysis on Euclidean Spaces, Princeton Univ. Press, Princeton N.J. 1971.

[98] SUNOUCHI, G., Saturation in the theory of best approximation, in: On Approximation Theory (P.L. Butzer - J. Korevaar, Eds.), pp. 72 - 88, ISNM 5, Birkhäuser, Basel 1964.

[99] SUNOUCHI, G. - C. WATARI, On determination of the class of saturation in the theory of approximation of functions, I Proc. Japan Acad. 34 (1958), 477 - 481; II Tôhoku Math.J. 11 (1959), 480 - 488.

[100] SZEGÖ, G., Orthogonal Polynomials, Amer. Math. Soc. Colloq. Publ. 23, Amer. Math. Soc., Providence, Rhode Island 1939.

[101] TAMARKIN, J.D., On integrable solutions of Abel's integral equation, Ann. of Math. 31 (1930), 219 - 229.

[102] TREBELS, W., Über Jackson- und Bernstein-Typ Ungleichungen für lineare Approximationsverfahren in $L^p(E_n)$ (to appear in Manuscripta Math.)

[103] TREBELS, W., On a Fourier-$L^1(E_n)$-multiplier criterium (to appear in Acta Sci. Math. (Szeged)).

[104] TURECKIĬ, A.H., Saturation classes in a space C (Russ.), Izv. Akad. Nauk SSSR Ser. Mat. 25 (1961), 411 - 442.

[105] TURNER, R.E.L., Eigenfunction expansions in Banach spaces, Quart.J.Math. Oxford Ser. 2 19 (1968), 193 - 211.

[106] WEYL, H., Bemerkungen zum Begriff des Differentialquotienten gebrochener Ordnung, Vierteljschr. Naturforsch. Ges. Zürich 62 (1917), 296 - 302.

[107] WING, G.M., The mean convergence of orthogonal series, Amer.J. Math. 72 (1950), 792 - 808.

[108] ZELLER, K. - W. BEEKMANN, Theorie der Limitierungsverfahren, Springer, Berlin 1970.

[109] ŽUK, V.V., Approximation of a 2π-periodic function by values of a certain bounded semiadditive operator II (Russ.), Vestnik Leningrad.Univ. 22 (1967), no 13, 41 - 50.

[110] ZYGMUND, A., Trigonometric Series I, II, Cambridge at the Univ. Press, Cambridge (2nd ed.) 1968.

LIST OF SYMBOLS

Symbol	*Description*	*page*
A_n^{α}	binomial coefficient	18
$AC_{loc}(0,\infty)$	set of locally absolutely continuous functions	24
$b_{\alpha}(x)$	fractional Riemann-Liouville integral	41
B^{ψ}	closed operator on X^{ψ}	11
$bv_{\alpha+1}$	subset of s	20
$BV_{\alpha+1}$	subset of $C_o[0,\infty)$ for $\alpha > 0$	24 36
$BV_{loc}(0,\infty)$	set of functions locally of bounded variation	24
$\mathbb{C}$	set of all complex numbers	9
(C^{α})	condition upon X and $\{P_k\}$	19
$(C,\alpha)_n$	Cesàro mean operator	18
(F)	first saturation condition	12
(F^*)	second saturation condition	14
$F[X;\mathcal{T}]$	Favard or saturation class of $\mathcal{T}$	6
I_{ω}^{α}	fractional integral operator	30
J^{α}	fractional Weyl integral operator	30
$\mathbb{K}$	subset of $\mathbb{P}$	12
$\mathcal{L}_{\Phi,\beta}$, $L_{\Phi,\beta}(\rho)$	Bessel potentials	63
$M = M(X;\{P_k\})$	set of all multipliers	10
$\mathbb{N}$	set of all positive integers	9
$\mathbb{P}$	set of all non-negative integers	1
P_k	projections	9
$\mathbb{R}$	set of all real numbers	9
$\mathcal{R}_{1,\beta}$, $R_{1,\beta}(\rho)$	Riesz means	42
$\mathcal{R}_{\Phi,\beta}$, $R_{\Phi,\beta}(\rho)$	general Riesz means	57
s	set of one-way sequences	10
$\mathcal{T}$	family of operators in [X]	3
$\mathcal{V}$, V_n	de La Vallée-Poussin means	67

$\mathfrak{M}_\Phi$, $W_\Phi(\rho)$	Abel-Cartwright means	54
X	Banach space	1
$[X]$	set of bounded linear operators from X into X	3
$[X]_M$	set of multiplier operators in $[X]$	10
X^ψ	subspace of X	11
$Y \subset X$	Y is continuously imbedded in X	4
$Y^{\sim X}$	completion of Y relative to X	14
$\mathbb{Z}$	set of all integers	9
$[\alpha]$	largest integer less than or equal to $\alpha \in \mathbb{R}$	9
Δ^β	fractional difference operator	20
$\oplus$	direct sum	51
$\lvert\cdot\rvert_\psi$	semi-norm on X	11
$\lvert\cdot\rvert_{\psi^\sim}$	semi-norm on $(X^\psi)^{\sim X}$	15
$\sim$	equivalence of two semi-norms	15
$\approx$	equivalence of two processes	7

Lecture Notes in Mathematics

Comprehensive leaflet on request

Vol. 146: A. B. Altman and S. Kleiman, Introduction to Grothendieck Duality Theory. II, 192 pages. 1970. DM 18,–

Vol. 147: D. E. Dobbs, Cech Cohomological Dimensions for Commutative Rings. VI, 176 pages. 1970. DM 16,–

Vol. 148: R. Azencott, Espaces de Poisson des Groupes Localement Compacts. IX, 141 pages. 1970. DM 16,–

Vol. 149: R. G. Swan and E. G. Evans, K-Theory of Finite Groups and Orders. IV, 237 pages. 1970. DM 20,–

Vol. 150: Heyer, Dualität lokalkompakter Gruppen. XIII, 372 Seiten. 1970. DM 20,–

Vol. 151: M. Demazure et A. Grothendieck, Schémas en Groupes I. (SGA 3). XV, 562 pages. 1970. DM 24,–

Vol. 152: M. Demazure et A. Grothendieck, Schémas en Groupes II. (SGA 3). IX, 654 pages. 1970. DM 24,–

Vol. 153: M. Demazure et A. Grothendieck, Schémas en Groupes III. (SGA 3). VIII, 529 pages. 1970. DM 24,–

Vol. 154: A. Lascoux et M. Berger, Variétés Kähleriennes Compactes. VII, 83 pages. 1970. DM 16,–

Vol. 155: Several Complex Variables I, Maryland 1970. Edited by J. Horváth. IV, 214 pages. 1970. DM 18,–

Vol. 156: R. Hartshorne, Ample Subvarieties of Algebraic Varieties. XIV, 256 pages. 1970. DM 20,–

Vol. 157: T. tom Dieck, K. H. Kamps und D. Puppe, Homotopietheorie. VI, 265 Seiten. 1970. DM 20,–

Vol. 158: T. G. Ostrom, Finite Translation Planes. IV. 112 pages. 1970. DM 16,–

Vol. 159: R. Ansorge und R. Hass. Konvergenz von Differenzenverfahren für lineare und nichtlineare Anfangswertaufgaben. VIII, 145 Seiten. 1970. DM 16,–

Vol. 160: L. Sucheston, Constributions to Ergodic Theory and Probability. VII, 277 pages. 1970. DM 20,–

Vol. 161: J. Stasheff, H-Spaces from a Homotopy Point of View. VI, 95 pages. 1970. DM 16,–

Vol. 162: Harish-Chandra and van Dijk, Harmonic Analysis on Reductive p-adic Groups. IV, 125 pages. 1970. DM 16,–

Vol. 163: P. Deligne, Equations Différentielles à Points Singuliers Reguliers. III, 133 pages. 1970. DM 16,–

Vol. 164: J. P. Ferrier, Seminaire sur les Algebres Complètes. II, 69 pages. 1970. DM 16,–

Vol. 165: J. M. Cohen, Stable Homotopy. V, 194 pages. 1970. DM 16,–

Vol. 166: A. J. Silberger, PGL_2 over the p-adics: its Representations, Spherical Functions, and Fourier Analysis. VII, 202 pages. 1970. DM 18,–

Vol. 167: Lavrentiev, Romanov and Vasiliev, Multidimensional Inverse Problems for Differential Equations. V, 59 pages. 1970. DM 16,–

Vol. 168: F. P. Peterson, The Steenrod Algebra and its Applications: A conference to Celebrate N. E. Steenrod's Sixtieth Birthday. VII, 317 pages. 1970. DM 22,–

Vol. 169: M. Raynaud, Anneaux Locaux Henséliens. V, 129 pages. 1970. DM 16,–

Vol. 170: Lectures in Modern Analysis and Applications III. Edited by C. T. Taam. VI, 213 pages. 1970. DM 18,–

Vol. 171: Set-Valued Mappings, Selections and Topological Properties of 2^X. Edited by W. M. Fleischman. X, 110 pages. 1970. DM 16,–

Vol. 172: Y.-T. Siu and G. Trautmann, Gap-Sheaves and Extension of Coherent Analytic Subsheaves. V, 172 pages. 1971. DM 16,–

Vol. 173: J. N. Mordeson and B. Vinograde, Structure of Arbitrary Purely Inseparable Extension Fields. IV, 138 pages. 1970. DM 16,–

Vol. 174: B. Iversen, Linear Determinants with Applications to the Picard Scheme of a Family of Algebraic Curves. VI, 69 pages. 1970. DM 16,–

Vol. 175: M. Brelot, On Topologies and Boundaries in Potential Theory. VI, 176 pages. 1971. DM 18,–

Vol. 176: H. Popp, Fundamentalgruppen algebraischer Mannigfaltigkeiten. IV, 154 Seiten. 1970. DM 16,–

Vol. 177: J. Lambek, Torsion Theories, Additive Semantics and Rings of Quotients. VI, 94 pages. 1971. DM 16,–

Vol. 178: Th. Bröcker und T. tom Dieck, Kobordismentheorie. XVI, 191 Seiten. 1970. DM 18,–

Vol. 179: Seminaire Bourbaki – vol. 1968/69. Exposés 347-363. IV. 295 pages. 1971. DM 22,–

Vol. 180: Séminaire Bourbaki – vol. 1969/70. Exposés 364-381. IV, 310 pages. 1971. DM 22,–

Vol. 181: F. DeMeyer and E. Ingraham, Separable Algebras over Commutative Rings. V, 157 pages. 1971. DM 16,–

Vol. 182: L. D. Baumert. Cyclic Difference Sets. VI, 166 pages. 1971. DM 16,–

Vol. 183: Analytic Theory of Differential Equations. Edited by P. F. Hsieh and A. W. J. Stoddart. VI, 225 pages. 1971. DM 20,–

Vol. 184: Symposium on Several Complex Variables, Park City, Utah, 1970. Edited by R. M. Brooks. V, 234 pages. 1971. DM 20,–

Vol. 185: Several Complex Variables II, Maryland 1970. Edited by J. Horváth. III, 287 pages. 1971. DM 24,–

Vol. 186: Recent Trends in Graph Theory. Edited by M. Capobianco/J. B. Frechen/M. Krolik. VI, 219 pages. 1971. DM 18,–

Vol. 187: H. S. Shapiro, Topics in Approximation Theory. VIII, 275 pages. 1971. DM 22,–

Vol. 188: Symposium on Semantics of Algorithmic Languages. Edited by E. Engeler. VI, 372 pages. 1971. DM 26,–

Vol. 189: A. Weil, Dirichlet Series and Automorphic Forms. V, 164 pages. 1971. DM 16,–

Vol. 190: Martingales. A Report on a Meeting at Oberwolfach, May 17-23, 1970. Edited by H. Dinges. V, 75 pages. 1971. DM 16,–

Vol. 191: Séminaire de Probabilités V. Edited by P. A. Meyer. IV, 372 pages. 1971. DM 26,–

Vol. 192: Proceedings of Liverpool Singularities – Symposium I. Edited by C. T. C. Wall. V, 319 pages. 1971. DM 24,–

Vol. 193: Symposium on the Theory of Numerical Analysis. Edited by J. Ll. Morris. VI, 152 pages. 1971. DM 16,–

Vol. 194: M. Berger, P. Gauduchon et E. Mazet. Le Spectre d'une Variété Riemannienne. VII, 251 pages. 1971. DM 22,–

Vol. 195: Reports of the Midwest Category Seminar V. Edited by J.W. Gray and S. Mac Lane. III, 255 pages. 1971. DM 22,–

Vol. 196: H-spaces – Neuchâtel (Suisse)- Août 1970. Edited by F. Sigrist, V, 156 pages. 1971. DM 16,–

Vol. 197: Manifolds – Amsterdam 1970. Edited by N. H. Kuiper. V, 231 pages. 1971. DM 20,–

Vol. 198: M. Hervé, Analytic and Plurisubharmonic Functions in Finite and Infinite Dimensional Spaces. VI, 90 pages. 1971. DM 16,–

Vol. 199: Ch. J. Mozzochi, On the Pointwise Convergence of Fourier Series. VII, 87 pages. 1971. DM 16,–

Vol. 200: U. Neri, Singular Integrals. VII, 272 pages. 1971. DM 22,–

Vol. 201: J. H. van Lint, Coding Theory. VII, 136 pages. 1971. DM 16,–

Vol. 202: J. Benedetto, Harmonic Analysis on Totally Disconnected Sets. VIII, 261 pages. 1971. DM 22,–

Vol. 203: D. Knutson, Algebraic Spaces. VI, 261 pages. 1971. DM 22,–

Vol. 204: A. Zygmund, Intégrales Singulières. IV, 53 pages. 1971. DM 16,–

Vol. 205: Séminaire Pierre Lelong (Analyse) Année 1970. VI, 243 pages. 1971. DM 20,–

Vol. 206: Symposium on Differential Equations and Dynamical Systems. Edited by D. Chillingworth. XI, 173 pages. 1971. DM 16,–

Vol. 207: L. Bernstein, The Jacobi-Perron Algorithm – Its Theory and Application. IV, 161 pages. 1971. DM 16,–

Vol. 208: A. Grothendieck and J. P. Murre, The Tame Fundamental Group of a Formal Neighbourhood of a Divisor with Normal Crossings on a Scheme. VIII, 133 pages. 1971. DM 16,–

Vol. 209: Proceedings of Liverpool Singularities Symposium II. Edited by C. T. C. Wall. V, 280 pages. 1971. DM 22,–

Vol. 210: M. Eichler, Projective Varieties and Modular Forms. III, 118 pages. 1971. DM 16,–

Vol. 211: Théorie des Matroïdes. Edité par C. P. Bruter. III, 108 pages. 1971. DM 16,–

Please turn over

Vol. 212: B. Scarpellini, Proof Theory and Intuitionistic Systems. VII, 291 pages. 1971. DM 24,-

Vol. 213: H. Hogbe-Nlend, Théorie des Bornologies et Applications. V, 168 pages. 1971. DM 18,-

Vol. 214: M. Smorodinsky, Ergodic Theory, Entropy. V, 64 pages. 1971. DM 16,-

Vol. 215: P. Antonelli, D. Burghelea and P. J. Kahn, The Concordance-Homotopy Groups of Geometric Automorphism Groups. X, 140 pages. 1971. DM 16,-

Vol. 216: H. Maaß, Siegel's Modular Forms and Dirichlet Series. VII, 328 pages. 1971. DM 20,-

Vol. 217: T. J. Jech, Lectures in Set Theory with Particular Emphasis on the Method of Forcing. V, 137 pages. 1971. DM 16,-

Vol. 218: C. P. Schnorr, Zufälligkeit und Wahrscheinlichkeit. IV, 212 Seiten 1971. DM 20,-

Vol. 219: N. L. Alling and N. Greenleaf, Foundations of the Theory of Klein Surfaces. IX, 117 pages. 1971. DM 16,-

Vol. 220: W. A. Coppel, Disconjugacy. V, 148 pages. 1971. DM 16,-

Vol. 221: P. Gabriel und F. Ulmer, Lokal präsentierbare Kategorien. V, 200 Seiten. 1971. DM 18,-

Vol. 222: C. Meghea, Compactification des Espaces Harmoniques. III, 108 pages. 1971. DM 16,-

Vol. 223: U. Felgner, Models of ZF-Set Theory. VI, 173 pages. 1971. DM 16,-

Vol. 224: Revètements Etales et Groupe Fondamental. (SGA 1). Dirigé par A. Grothendieck XXII, 447 pages. 1971. DM 30,-

Vol. 225: Théorie des Intersections et Théorème de Riemann-Roch. (SGA 6). Dirigé par P. Berthelot, A. Grothendieck et L. Illusie. XII, 700 pages. 1971. DM 40,-

Vol. 226: Seminar on Potential Theory, II. Edited by H. Bauer. IV, 170 pages. 1971. DM 18,-

Vol. 227: H. L. Montgomery, Topics in Multiplicative Number Theory. IX, 178 pages. 1971. DM 18,-

Vol. 228: Conference on Applications of Numerical Analysis. Edited by J. Ll. Morris. X, 358 pages. 1971. DM 26,-

Vol. 229: J. Väisälä, Lectures on n-Dimensional Quasiconformal Mappings. XIV, 144 pages. 1971. DM 16,-

Vol. 230: L. Waelbroeck, Topological Vector Spaces and Algebras. VII, 158 pages. 1971. DM 16,-

Vol. 231: H. Reiter, L^1-Algebras and Segal Algebras. XI, 113 pages. 1971. DM 16,-

Vol. 232: T. H. Ganelius, Tauberian Remainder Theorems. VI, 75 pages. 1971. DM 16,-

Vol. 233: C. P. Tsokos and W. J. Padgett. Random Integral Equations with Applications to Stochastic Systems. VII, 174 pages. 1971. DM 18,-

Vol. 234: A. Andreotti and W. Stoll. Analytic and Algebraic Dependence of Meromorphic Functions. III, 390 pages. 1971. DM 26,-

Vol. 235: Global Differentiable Dynamics. Edited by O. Hájek, A. J. Lohwater, and R. McCann. X, 140 pages. 1971. DM 16,-

Vol. 236: M. Barr, P. A. Grillet, and D. H. van Osdol. Exact Categories and Categories of Sheaves. VII, 239 pages. 1971, DM 20,-

Vol. 237: B. Stenström. Rings and Modules of Quotients. VII, 136 pages. 1971. DM 16,-

Vol. 238: Der kanonische Modul eines Cohen-Macaulay-Rings. Herausgegeben von Jürgen Herzog und Ernst Kunz. VI, 103 Seiten. 1971. DM 16,-

Vol. 239: L. Illusie, Complexe Cotangent et Déformations I. XV, 355 pages. 1971. DM 26,-

Vol. 240: A. Kerber, Representations of Permutation Groups I. VII, 192 pages. 1971. DM 18,-

Vol. 241: S. Kaneyuki, Homogeneous Bounded Domains and Siegel Domains. V, 89 pages. 1971. DM 16,-

Vol. 242: R. R. Coifman et G. Weiss, Analyse Harmonique Non-Commutative sur Certains Espaces. V, 160 pages. 1971. DM 16,-

Vol. 243: Japan-United States Seminar on Ordinary Differential and Functional Equations. Edited by M. Urabe. VIII, 332 pages. 1971. DM 26,-

Vol. 244: Séminaire Bourbaki - vol. 1970/71. Exposés 382-399. IV, 356 pages. 1971. DM 26,-

Vol. 245: D. E. Cohen, Groups of Cohomological Dimension One. V, 99 pages. 1972. DM 16,-

Vol. 246: Lectures on Rings and Modules. Tulane University Ring and Operator Theory Year, 1970-1971. Volume I. X, 661 pages. 1972. DM 40,-

Vol. 247: Lectures on Operator Algebras. Tulane University Ring and Operator Theory Year, 1970-1971. Volume II. XI, 786 pages. 1972. DM 40,-

Vol. 248: Lectures on the Applications of Sheaves to Ring Theory. Tulane University Ring and Operator Theory Year, 1970-1971. Volume III. VIII, 315 pages. 1971. DM 26,-

Vol. 249: Symposium on Algebraic Topology. Edited by P. J. Hilton. VII, 111 pages. 1971. DM 16,-

Vol. 250: B. Jónsson, Topics in Universal Algebra. VI, 220 pages. 1972. DM 20,-

Vol. 251: The Theory of Arithmetic Functions. Edited by A. A. Gioia and D. L. Goldsmith VI, 287 pages. 1972. DM 24,-

Vol. 252: D. A. Stone, Stratified Polyhedra. IX, 193 pages. 1972. DM 18,-

Vol. 253: V. Komkov, Optimal Control Theory for the Damping of Vibrations of Simple Elastic Systems. V, 240 pages. 1972. DM 20,-

Vol. 254: C. U. Jensen, Les Foncteurs Dérivés de $\varprojlim$ et leurs Applications en Théorie des Modules. V, 103 pages. 1972. DM 16,-

Vol. 255: Conference in Mathematical Logic - London '70. Edited by W. Hodges. VIII, 351 pages. 1972. DM 26,-

Vol. 256: C. A. Berenstein and M. A. Dostal, Analytically Uniform Spaces and their Applications to Convolution Equations. VII, 130 pages. 1972. DM 16,-

Vol. 257: R. B. Holmes, A Course on Optimization and Best Approximation. VIII, 233 pages. 1972. DM 20,-

Vol. 258: Séminaire de Probabilités VI. Edited by P. A. Meyer. VI, 253 pages. 1972. DM 22,-

Vol. 259: N. Moulis, Structures de Fredholm sur les Variétés Hilbertiennes. V, 123 pages. 1972. DM 16,-

Vol. 260: R. Godement and H. Jacquet, Zeta Functions of Simple Algebras. IX, 188 pages. 1972. DM 18,-

Vol. 261: A. Guichardet, Symmetric Hilbert Spaces and Related Topics. V, 197 pages. 1972. DM 18,-

Vol. 262: H. G. Zimmer, Computational Problems, Methods, and Results in Algebraic Number Theory. V, 103 pages. 1972. DM 16,-

Vol. 263: T. Parthasarathy, Selection Theorems and their Applications. VII, 101 pages. 1972. DM 16,-

Vol. 264: W. Messing, The Crystals Associated to Barsotti-Tate Groups: with Applications to Abelian Schemes. III, 190 pages. 1972. DM 18,-

Vol. 265: N. Saavedra Rivano, Catégories Tannakiennes. II, 418 pages. 1972. DM 26,-

Vol. 266: Conference on Harmonic Analysis. Edited by D. Gulick and R. L. Lipsman. VI, 323 pages. 1972. DM 24,-

Vol. 267: Numerische Lösung nichtlinearer partieller Differential- und Integro-Differentialgleichungen. Herausgegeben von R. Ansorge und W. Törnig, VI, 339 Seiten. 1972. DM 26,-

Vol. 268: C. G. Simader, On Dirichlet's Boundary Value Problem. IV, 238 pages. 1972. DM 20,-

Vol. 269: Théorie des Topos et Cohomologie Etale des Schémas. (SGA 4). Dirigé par M. Artin, A. Grothendieck et J. L. Verdier. XIX, 525 pages. 1972. DM 50,-

Vol. 270: Théorie des Topos et Cohomologie Etle des Schémas. Tome 2. (SGA 4). Dirige par M. Artin, A. Grothendieck et J. L. Verdier. V, 418 pages. 1972. DM 50,-

Vol. 271: J. P. May, The Geometry of Iterated Loop Spaces. IX, 175 pages. 1972. DM 18,-

Vol. 272: K. R. Parthasarathy and K. Schmidt, Positive Definite Kernels, Continuous Tensor Products, and Central Limit Theorems of Probability Theory. VI, 107 pages. 1972. DM 16,-

Vol. 273: U. Seip, Kompakt erzeugte Vektorräume und Analysis. IX, 119 Seiten. 1972. DM 16,-

Vol. 274: Toposes, Algebraic Geometry and Logic. Edited by. F. W. Lawvere. VI, 189 pages. 1972. DM 18,-

Vol. 275: Séminaire Pierre Lelong (Analyse) Année 1970-1971. VI, 181 pages. 1972. DM 18,-

Vol. 276: A. Borel, Représentations de Groupes Localement Compacts. V, 98 pages. 1972. DM 16,-

Vol. 277: Séminaire Banach. Edité par C. Houzel. VII, 229 pages. 1972. DM 20,-